The Psoas Book

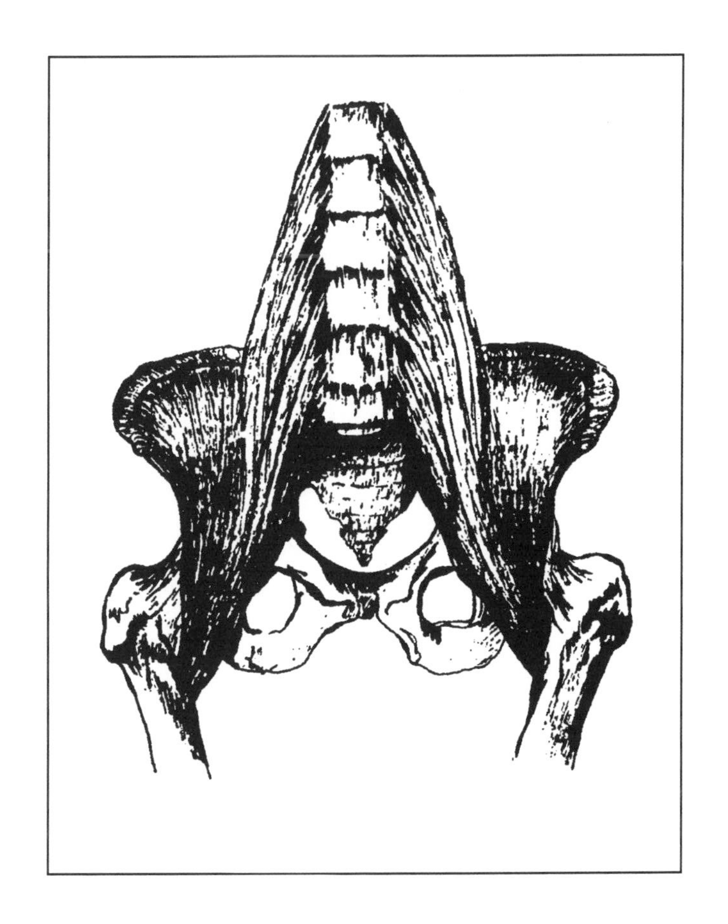

by Liz Koch

The Psoas Book

second edition

1997

For information and requests contact:
Guinea Pig Publications
1226 P.O. Box
Felton, CA. 95018

Library of Congress Cataloging-in-Publication Data
1. Koch, Liz
2. The Psoas Book
3. 1. Health. 2. Self-Help. 3. Women
4. 97 - 71684
5.ISBN 0-9657944-0-7

Second Printing 1997 Third Printing 1999 Fourth Printing 2001

Dedicated to my mother and father, who I have grown to truly appreciate. Without their support this second edition would not of been possible.

What People Are Saying about the Psoas Workshops

"Liz provides this important muscle with an informed and experiential context that moves the mystique of the psoas into the realm of core experience and the opportunity for self-healing"
Ronnie Oliver, Aston-Patterning Practitioner and Teacher (San Francisco, California)

"The workshop is very worthwhile for ALL practitioners of health and spirituality. I loved it!"
Kathi Flanagan,Yoga Teacher (Tampa, Florida)

"After the workshop my dancing changed...I had taken a quantum leap. One choreographer said my whole pelvis had '"opened" and my dancing had reached a higher level."
Elaine Nakashima, Dancer/Therapist (Santa Barbara, California)

The workshop answered so many of my questions...light bulbs went on all over my body."
Shivan Sarna, Yoga Teacher (Sarasota, Florida)

"I am fully aware of my psoas now and how vital it is. I appreciate it's movement in my voice work"
Workshop Participant, Voice & Theater Student (Boston, Massachusetts)

"After attending the workshop I found I was more focused on the subtleties of groin and abdominal work with my rolfing sessions. I have been using the exercises as I work with clients and sometimes teaching them these subtle movements to continue as exercises at home"
Janis G. Davis M.D., Rolfer for 25 years (St. Petersburg, Florida)

The Psoas Book

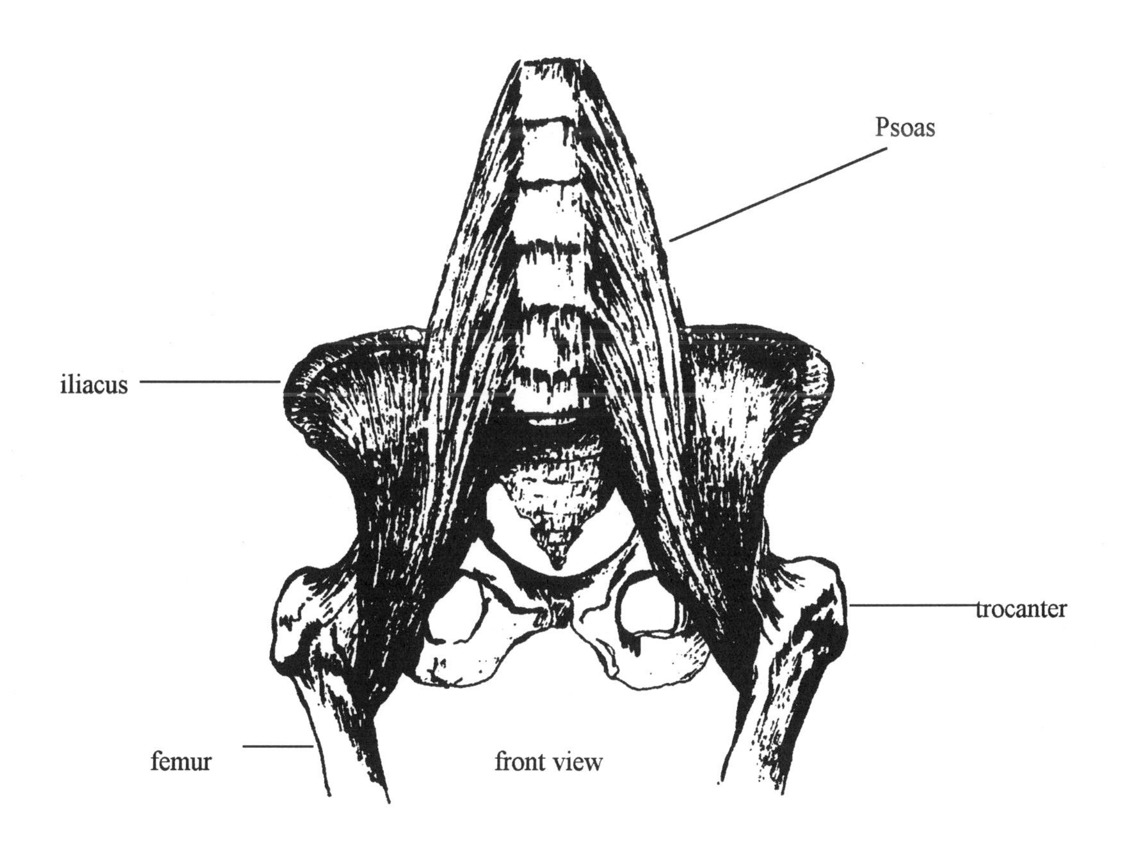

by Liz Koch

What People Are Saying about the Psoas Workshops

Liz's specific attention on the Psoas muscle is brilliant. In vibrational and fluid movement bodywork, the psoas is key. The more fluid we become, the freer the psoas must be!
Patricia Cramer, Founder of The World School of Massage, (San Francisco, California)

I loved the workshop...I realized that I could feel the psoas!
Participant, (Santa Barbara, California)

"I found the workshop deeply enlightening...a new awareness of what has been a stuck, or dense area of my body."
Participant (San Francisco, California)

"I have always suffered from menstrual pain, but learning to release the psoas muscle has provided me with a totally natural method of relief that works miraculously"
Participant (San Francisco, California)

"What I especially liked about the workshop was the breadth of information, the variety of experience, and the depth of knowledge."
Suki Munsell Ph.D., Founder and Director of Dynamic Health and Fitness Institute (Corte Madera, California)

Liz combines an eclectic knowledge with a working knowledge of the psoas, creating a fascinating and well-rounded approach."
Liz Bruno, Dance Instructor, (San Francisco, California)

Liz Koch awakened my interest in the psoas many years ago. The amount of healing this information has yielded has been unfathomable. The psoas is "the" problem muscle.
Victor J. Collins, D.C., (Santa Cruz California)

"When one tugs at a single thing in nature, he finds it attached to the rest of the world."

John Muir

Contents

Introduction

The material presented in this booklet has been collected over the past seven years, with the intention of offering it to students who have requested reading material and information concerning the psoas muscle. I have collected here the information I have found most helpful in my search of balanced posture.

I have a history of scoliosis (spinal curvature), muscle spasms, limited range of movement, and general awkwardness-something that I assumed I would live with and work around for the rest of my life. It was not until my twenties that an introduction to yoga opened me to the possibilities of experiencing my body differently. It was a powerful experience, and I attempted to continue along those lines. Finding it too difficult to work alone, I sought and found my first teacher and beloved friend, Robert Cooley. I worked with Bob for three years at his Boston school, ***The Moving Center, Inc***. Originally a dancer, Bob had become interested in why dancers so easily injure themselves. The more he learned, the more he seemed to move away from dance (though he loves dancing) into a more specific interest in movement. As Bob's perceptive eye and keen sense of movement grew more experienced, he taught me what he knew, encouraging me to learn to sense my own body while developing an eye for seeing the human body in motion.

While Bob was investing his time and effort in releasing and lengthening his psoas, I spent the first six months that I worked with him just trying to find out what I was looking for, and where it was in my body. It took a long time to realize how subtle and deep the sensation could be. At first I felt miles way from myself; but as I worked quietly, and with Bob's encouragement and humor, I learned the value of directing my attention toward my sensation.

I found the territory unknown and often scary, and sometimes it felt all too personal. And yet being in that warm sunlit room, surrounded by green plants, oriental carpets, and friendly people, I allowed myself the time and opportunity to venture inside and look around.

We were all trying, in the same way, to speak of what we were sensing. "I sense my psoas on the left side more contracted than on the right...now I sense it stretching. I sense a quivering around my rib cage and a sensation of heat moving into my legs..." We followed our attention as it moved through the body. Later, it was fun trying to catch an emotion or a thought as it flew by connected to a sensation like a tail on a kite. "I have an ache sensation in my right hip socket, I notice a feeling of frustration and anger...an image of my father comes up, I think I'd like to kick him..." And then laughter - a sense of humor towards my conditioning was being cultivated.

When I left Boston for out west I was both ready and scared. I wondered what would happen without the support and understanding of Bob and the studio. Once settled in California I began again, on my own. Eventually I got the idea to try to gather other people to work with me. I wrote Bob at the same time that he wrote to me, suggesting that I start a class. And so I did. Four years went by as I continued to learn from teaching. Staying one step ahead of my students, I have often felt that I was being pushed by my the class, rather than leading it. Through teaching and studying, I explored various methods as I searched for more understanding of the process going on under my nose.

I met my acupressure teacher, Aminah Raheem in Santa Cruz, a year after starting to teach, and became interested in *Jin Shin Do*, a gentle meditative form of Japanese acupressure based on the

Chinese meridian point system. I became certified as a practitioner. Practicing *Jin Shin Do* opened to me a whole new way of perceiving the body, and this different perspective influenced my approach to teaching *"Postural Transformation"*.

I was not very touched by most of the methods I encountered (although appreciating certain aspects of them) until a birthday present gave me an opportunity to experience *Aston-Patterning,* an approach developed by Judith Aston to free the body of unnecessary stress, both structurally and in everyday movement. Generally speaking, I had approached my body directly; that is, I worked through my own awareness rather than seeking people to "work on me". In fact, I felt especially offended when anyone wanted to try to correct my posture. Methods of working were for me tools to develop my sensitivity and to increase my awareness; any postural change was a bonus.

However, I was moved by the work with Ronnie Neufeld-Oliver, a masseuse and *Aston-Patterning* teacher. She was equally interested in acupressure, giving us a unique opportunity to experience each other's way of working, to exchange ideas, and to develop a close friendship.

Aston-Patterning responded to a question I had begun to formulate concerning the need for an organized way of approaching the body. I wasn't sure yet what I meant by that, but I sensed that Bob's work was not formed into a method. I found *Patterning* responded to that need, not by imposing an image of right posture on my body, but by evoking my body's natural ability to organize, balance and coordinate itself. I also like the method because it works not only with releasing the psoas, but also with engaging it in movement. For the first time I experienced volume, depth, and a three-dimensional quality to my body. My scoliosis began unraveling, making visible its emotional and attitudinal components. This might be considered a miracle compared to the standard approach to scoliosis, which includes stringent exercise, back braces, drugs, body casts and vertebrae fusing. With *Patterning*, each step leads to the next, so that in a manner of speaking the work is always digestible. I feel an elaborate three-dimensional mosaic pattern is being constructed. The further I go toward sensing myself, the more I find my body opens to receive and experience life.

Although this booklet focuses on only one aspect of the body, the psoas muscle, I think it is impossible to separate out one element from the whole. Therefore I suggest that anyone pursuing an understanding of his or her body try to perceive and work with the body as an integrated whole, as a receiver, transformer, and transmitter of substances or energies, which functions as part of a larger universal body and which demands our attention, sensitivity, and respect.

Liz Koch
San Francisco, California
1981

Introduction

What began as a simple student manual, grew over the years to be a layperson and professional guide to the iliopsoas muscle. I have worked with the psoas muscle for over 20 years and still find it an intriguing subject. The psoas represents the deepest, instinctual qualities of energy in the human being. It is from the area of the psoas that wise women and wise men ground themselves. With an integrated , well functioning psoas comes a quiet, safe haven to move from and be within. The image that appears for me is a tribal image; a wise bushman, who's instinctual self has transcended survival skills to the "Fine Art of Being" on Mother Earth. From this deeply grounded, stable place we allow the heart and mind to soar. Only when the psoas is free to move, can the energy of the body flow smoothly, the emotions balance, and our thoughts be integrated.

I would like to acknowledge all the people who made this second edition possible. My son Adam Oberdorfer who taught me to use a computer and who has served as my computer graphics technician. My husband Jeff Oberdorfer and children Megan and Lily who "put up" with my constant dialogue about the psoas muscle. Victor Collins D.C. who has served as both healer and teacher, thank you for helping me articulate my ideas as well as my own psoas muscle! My colleague Ronnie Oliver who continues to be my best friend and with whom I love sharing ideas and combining our teaching skills. And Bob Cooley, without whom I might never have discovered I had a psoas muscle. He is still my teacher, good friend and greatest supporter of my work. Last but not least, I thank all my students who keep asking me questions for which I do not have the answers.

Liz Koch
Felton, California
1997

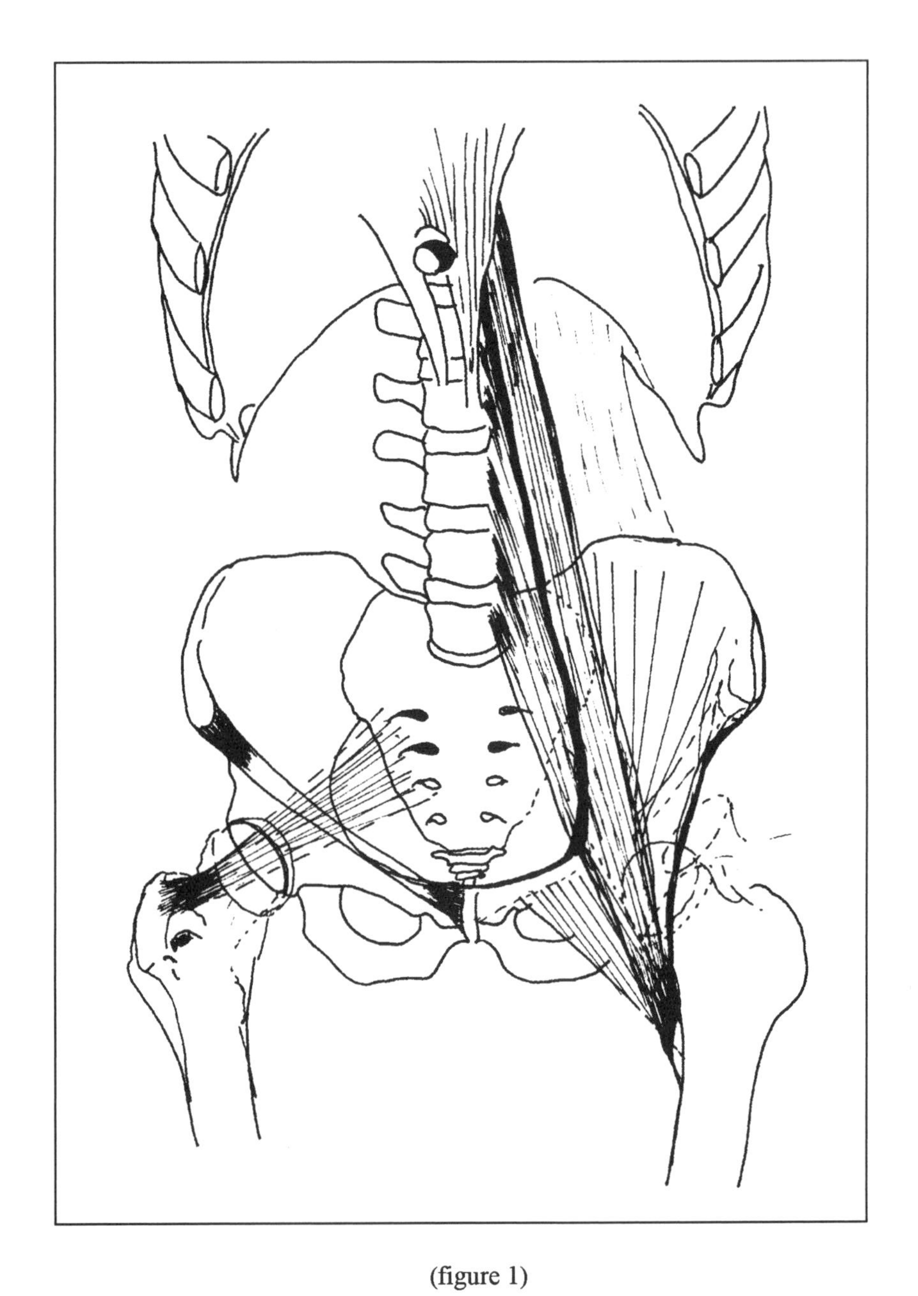

(figure 1)

Chapter One
Location

The Iliopsoas muscle or ***psoas muscle*** (pronounced so-az) is the keystone of a balanced, well-organized body. It is a massive muscle, approximately 16 inches long, that directly links the ribcage and trunk with the legs. There are two psoas muscles, one attaching on either side of the lumbar spine. Each consists of a psoas major and a psoas minor. (see figure 1) The psoas major has its' origin on the side of the body and transverse processes of the spinal vertebrae. It attaches at the twelfth thoracic vertebra (T12) and each of the five lumbar vertebrae (L 1-5). The psoas major passes through the pelvis, over the ball and socket of the hip joint and attaches at the inner side of the lesser trocanter of the femur bone.

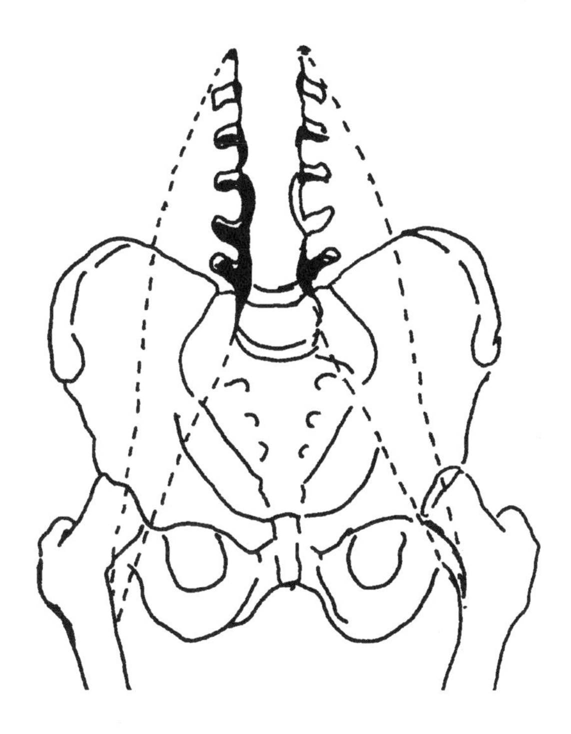

(figure 2 - front view)

Location

The psoas does not attach directly to the pelvis, (figure 2) but vitally influences it through its' relationship with the ribcage and femurs, and more importantly through its' attachment with the iliacus muscle by way of a common tendon at the lesser trocanter. (figure 3)

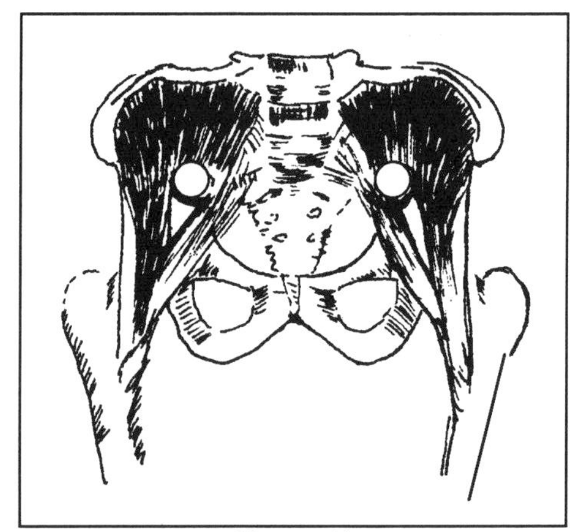

(figure 3)

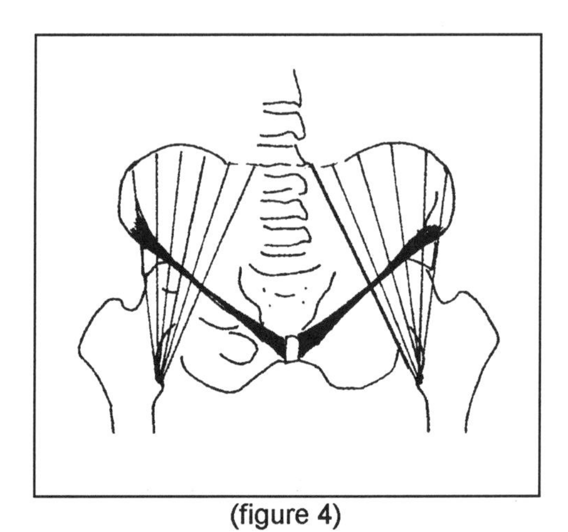

(figure 4)

The **iliacus** is a fan-shaped muscle lining the inside of the pelvic basin (figure 4). The iliacus and psoas are spoken of together as the **iliopsoas** muscle group. The tone of the psoas affects the pelvis and all its' contents by way of the iliacus muscle. The same is true of the iliacus and the functioning of the organs and viscera. They too may affect the health and vitality of the psoas muscle.

The psoas minor is a vanishing muscle. It attaches at T12 and appears to attach in a thin tendon at the pelvic rim. (see figure 1) A relic of our primordial ancestry it is thought to be a disappearing muscle as we evolve from a semi -flexed to an upright being. It is possible to have only one psoas minor (on one side), one on both sides or none at all.

The lumbar spine is located approximately in the center of the body. The psoas is experienced at the body's axis; that is, at its' deepest core. (Figure 5) Constricted back muscles may give us the sensation that a ridged muscular structure is what holds us up against the powerful force of gravity. Similarly, the knobby projections we feel along our back may lead us to believe that the spinal column is located in the back of the body. What we touch when we run our hands along our backs are the transverse processes, the wing-like protrusions of the spine. However, the actual body of the vertebra sit more toward the center or mid-line of the body than is usually perceived. The vertebrae of the lumbar spine, being the largest, lie at the deepest core of our structure. The spinal column is deeply set and occupies approximately one-half the diameter of the body from the front to the back. Its' central position, subtle triangular shape along with the bilateral symmetry give the spine its strength. The S-shaped curve (figure 6) configuration give it the ability to not only support the weight of the whole body when upright but to offer a resiliency to the forces of gravity.

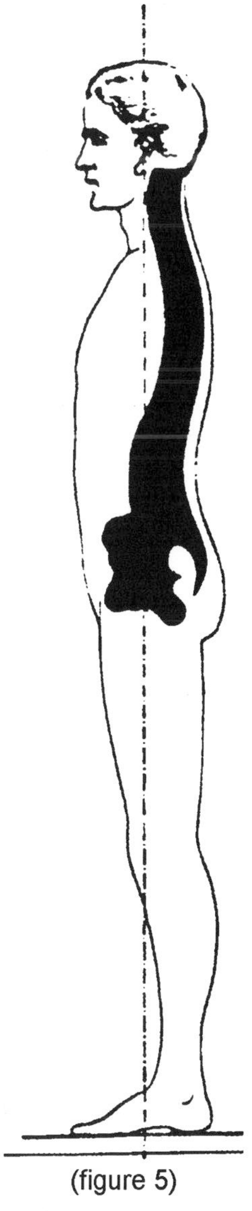

(figure 5)

Another common misconception is the location of the hip sockets. If asked where their hip sockets are, most people will point to the sides of their hips, because that is where movement most often is experienced

(figure 6)

when the leg is in action. Actually, that is the location of the greater trocanter. The ball and socket is located in the front of the pelvis where the socket of the pelvis and the ball of the femur come together, creating a joint. (figure 7) At the joints, weight is transferred from a single structure, the trunk, into the two legs. Movement of the legs takes place at the hip joints. Passing over the hip joints, the psoas muscle directly affects the range of movement and rotation in the pelvis and legs.(figures 9, 11)

The direction of the psoas through the body, is both vertical and diagonal (in 2 planes), moving from deep within at T12, and surfacing at the hip sockets; then attaching to the lessor trocanter. It gradually spreads diagonally as it flares over the hip sockets, as well as traversing diagonally through the pelvic cavity. (figure 11) The psoas is kinesthetically sensed as deepest at its' attachments, surfacing as it passes over the hip sockets. The other diagonal is referred to as the *psoatic shelf*. The psoas as a shelf provides a muscular support for the internal organs. (figure 8)

A clear image of the muscles of the femur and their attachments to the pelvis helps to give a better idea of the location of the psoas in relation to the legs. As shown in (figure 13) the thigh muscles have thier origins at the pelvis To complete this picture of the pelvis as a foundation, and the psoas as the keystone of that foundation, look at the large muscles of the arms shoulders, and back, noting that they too attach to the pelvis (figure 12 A & B).

The lumbodorsal junction is a particularly important intersection. It is here that the psoas first inserts into the spine, behind the diaphragm and on the sides of the twelfth dorsal vertebra (T12), and begins its' journey downward toward the legs. On the back of T12, the massive trapezius muscle attaches and moves upward towards the skull.(figure 12 A & B))

The action of the diaphragm, a lordotic muscle, moves it like a piston through the trunk. The diaphragm is not generally experienced as part of the lumbar musculature, and yet its' attachments reach as far as the 4th and 5th vertebrae of the lumbar spine. (figure 1)

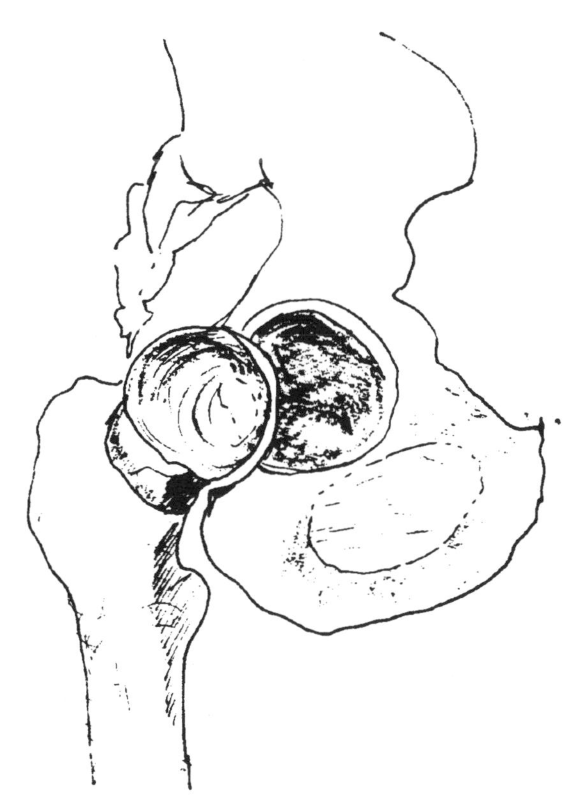

(figure 7 -A)

The diaphragm is a muscular structure seperating the chest from the abdomen. Forming a floor for the thoracic cage the heart rests on the top of the diaphragm. As a roof above the abdominal cavity the liver, stomach and spleen are immediately beneath its' under surface.

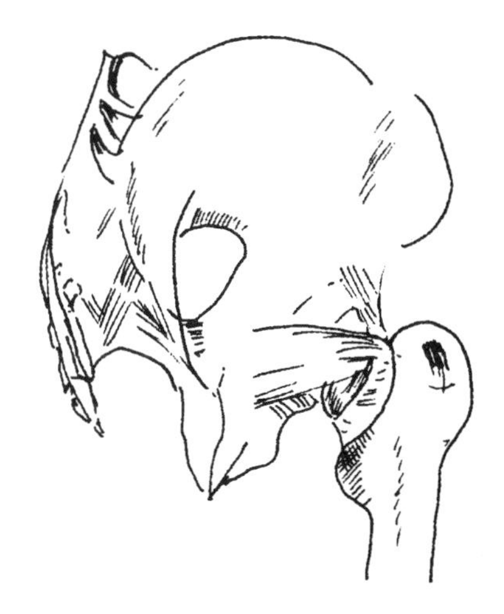

(figure 7-B)

(figure 10 A & B) All of the viscera are in close contact with the diaphragm and are directly connected with its' tissues. Like the psoas muscle the diaphragm is positioned

Location

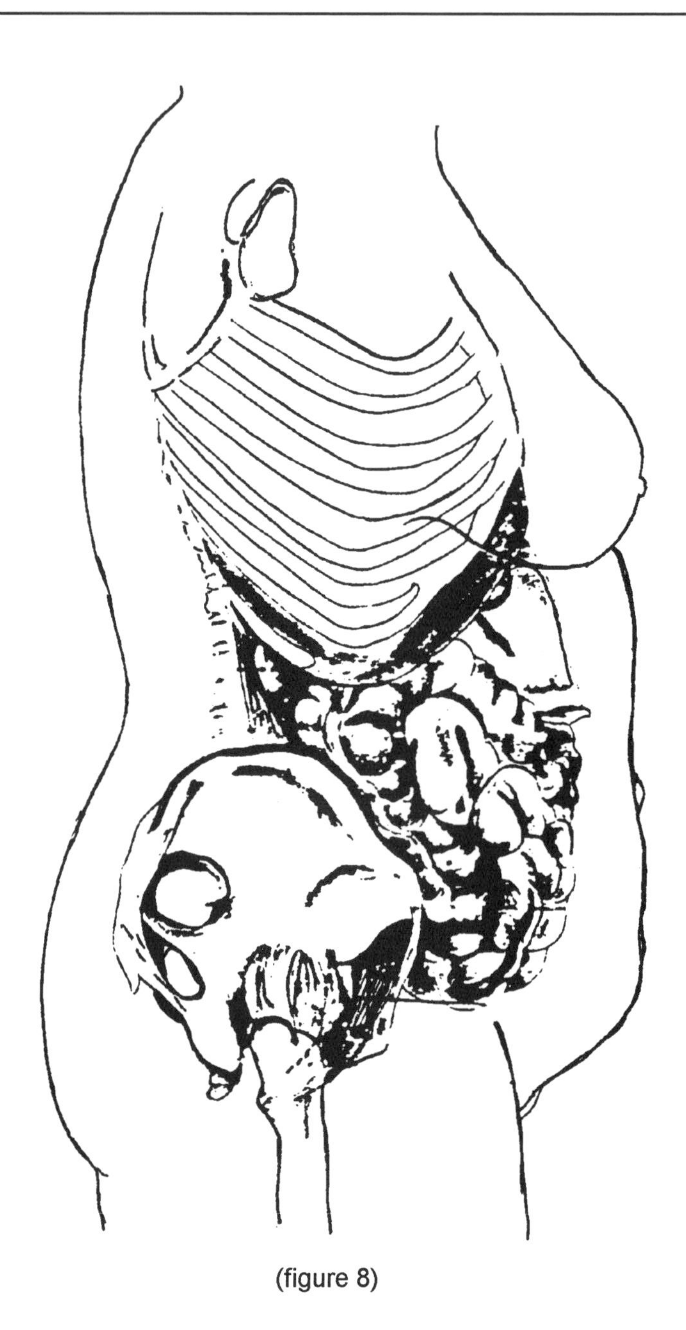

(figure 8)

such that it responds and is affected by two different rhythms of the body; the visceral and the skeletal. As it moves up and down through the body, it massages not only the organs but the vertebrae moving **synovial fluid** through the spine to the brain.

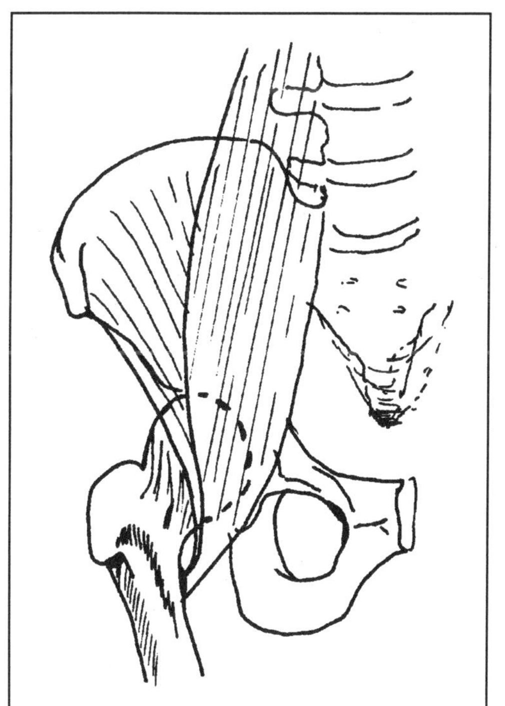

(figure 9 -A)

The esophagus and nerves actually penetrate the diaphragm. The major aorta (artery) from the heart lies just behind it. The psoas and aorta both move through the trunk and seem to follow each other as they pass over the hip joints, where the aorta, having already split into two arteries just below the navel, moves into and through the legs.(figure 21)

The lumbar nerve plexus is a complex network of nerves moving through and around the psoas muscle. (figure 23) Many of the nerves are actually embedded in its' surface. A complex communication to and from the viscera and the brain involves the psoas muscle. The psoas responds or interprets messages between one and the other by torquing specific vertebrae.

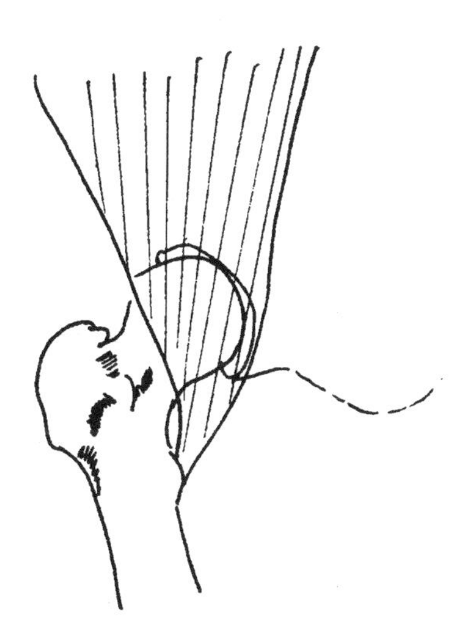

(figure 9-B)

The nerve ganglia of the lumbar may be thought of as our abdominal brain. The psoas resides where our gut feelings are felt. The upper psoas and diaphragm meet at the junction known as the *solar plexus*. The solar plexus is energetically a center for personal power and the control of feelings. Ida Rolf in her book **Rolfing - The Integration of the Human Structure** explains that *"along the anterior lateral surface of the entire spine lies the sympathetic trunk of the autonomic nervous system...This more archaic nerve unit is thought to be at a level below our voluntary control, although recent findings shed a certain doubt on this. It forms a series of ganglia that are centers integrating associated nerve elements. The great solar plexus, locus of the largest of these, is sometimes called the abdominal brain. It lies approximately at the level where the psoas and diaphragm juxtapose. The lumbar plexus, with its' network for visceral and muscular intercommunication, is the next lower neighbor and is embedded in the surface of the psoas itself."* (11 pg. 113)

Each kidney is situated to the side of the psoas muscle. The bladder, viscera and reproductive organs lie in front of the psoas (figure 24)

It is the vital and dynamic interrelationship of the psoas with the diaphragm, organs, blood, and nerves that gives the psoas muscle a powerful unifying function.

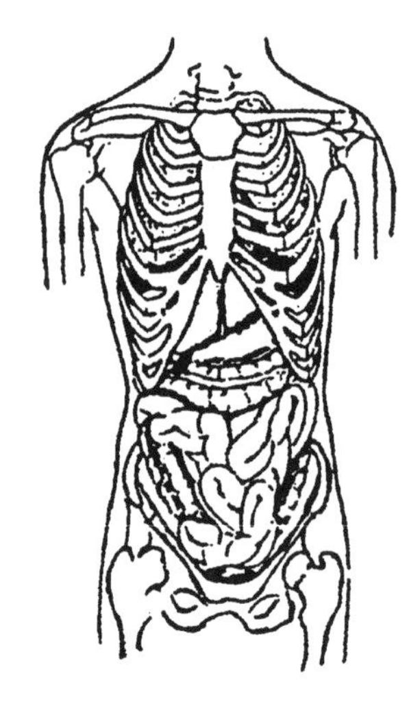

(figure 10 A)

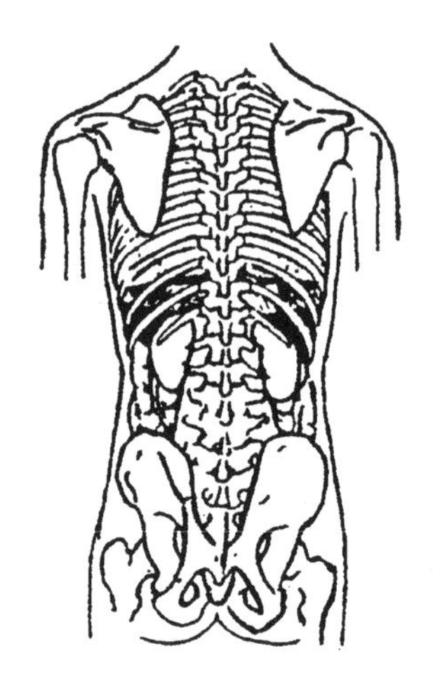

(figure 10 B)

Location

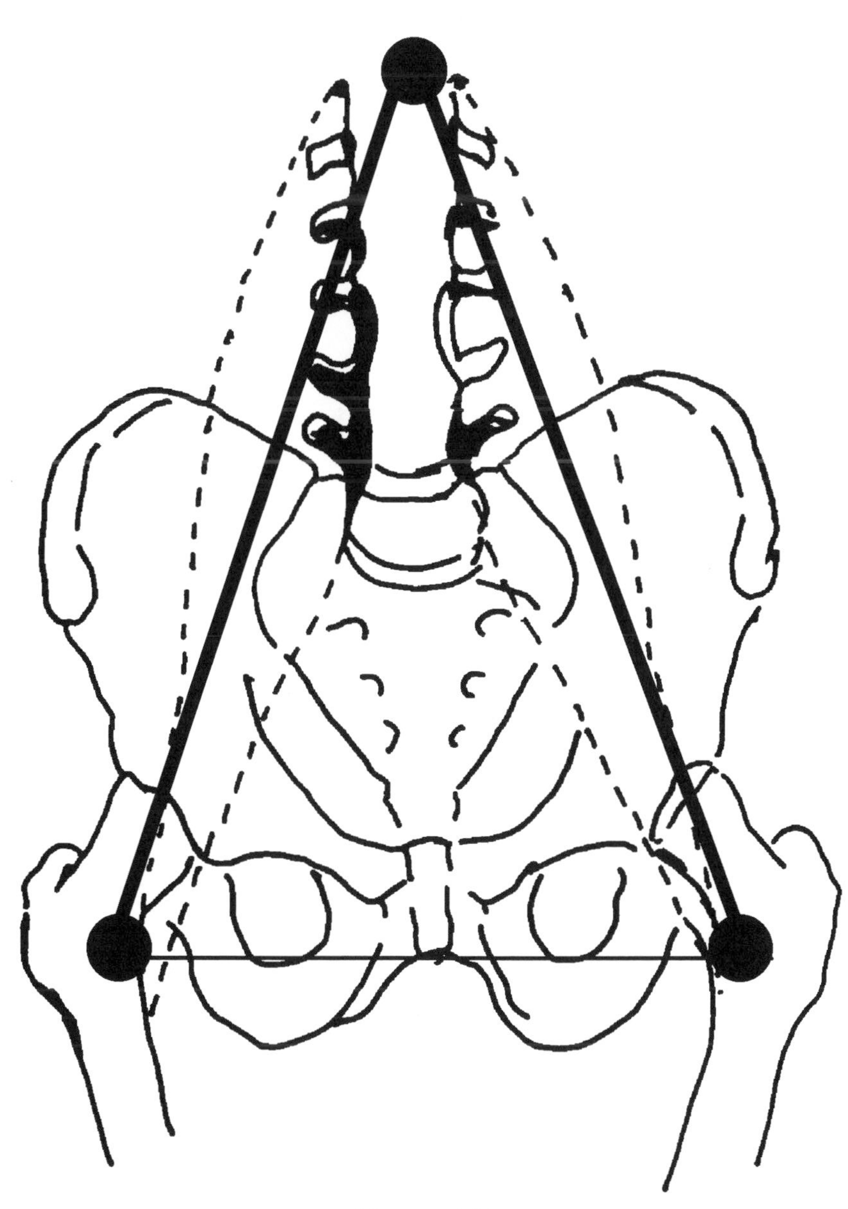

(figure 11)

Location

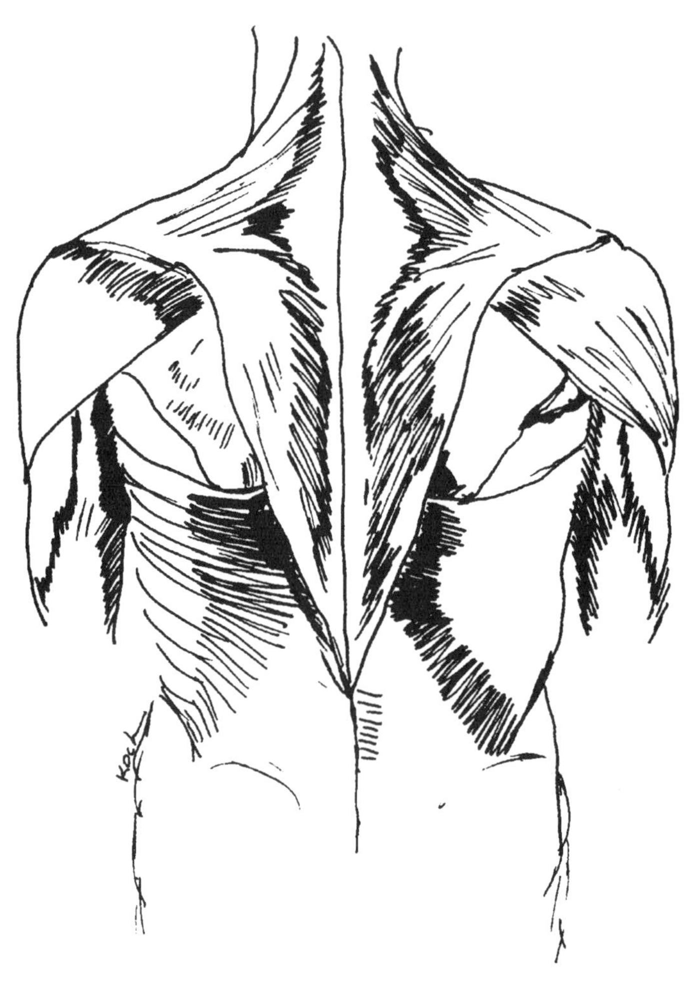

(figure 12 -A)

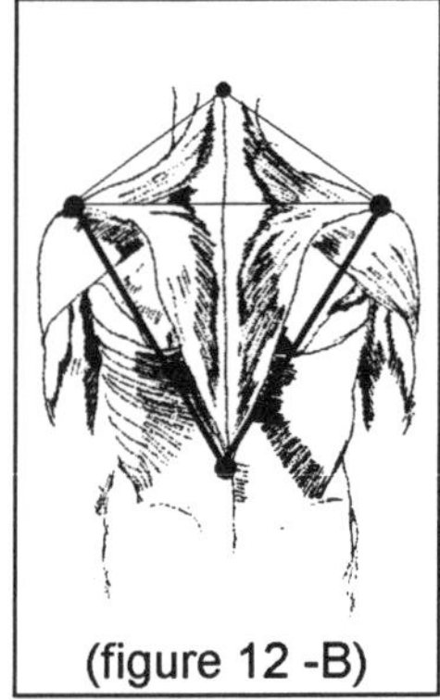

(figure 12 -B)

Location

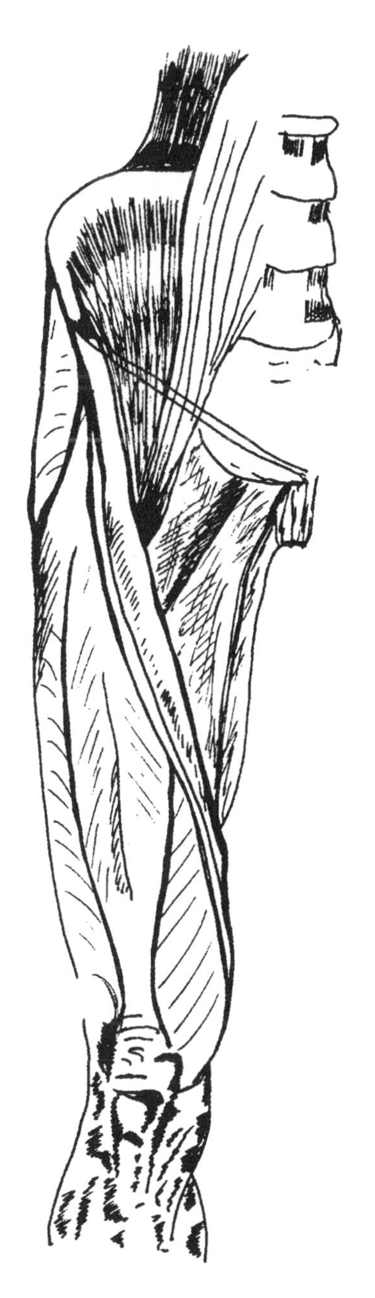

(figure 13-A)

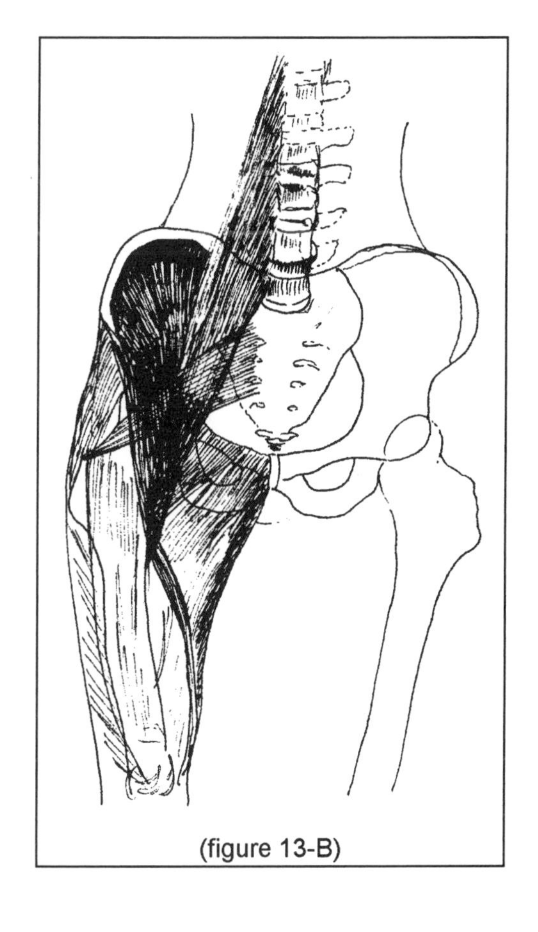

(figure 13-B)

Location

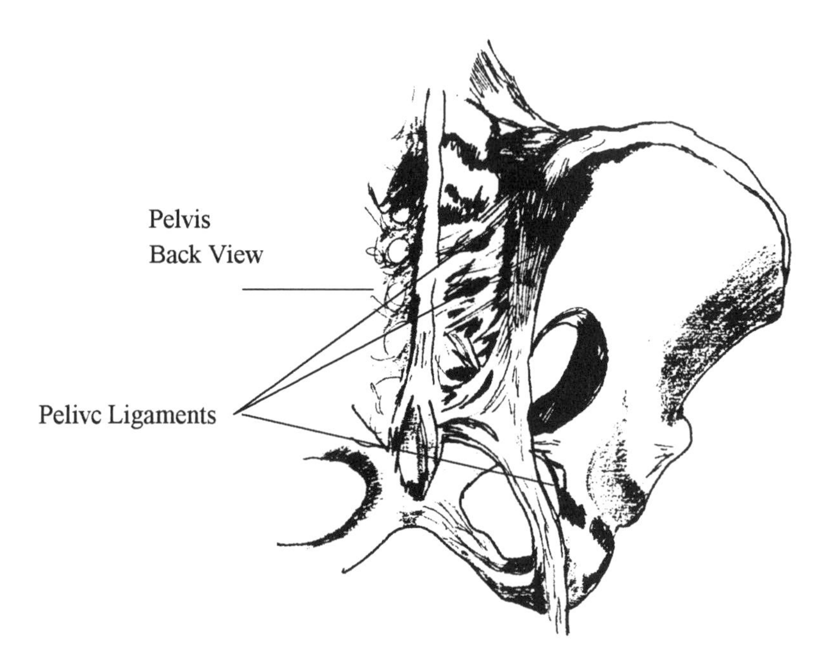

(figure 14)

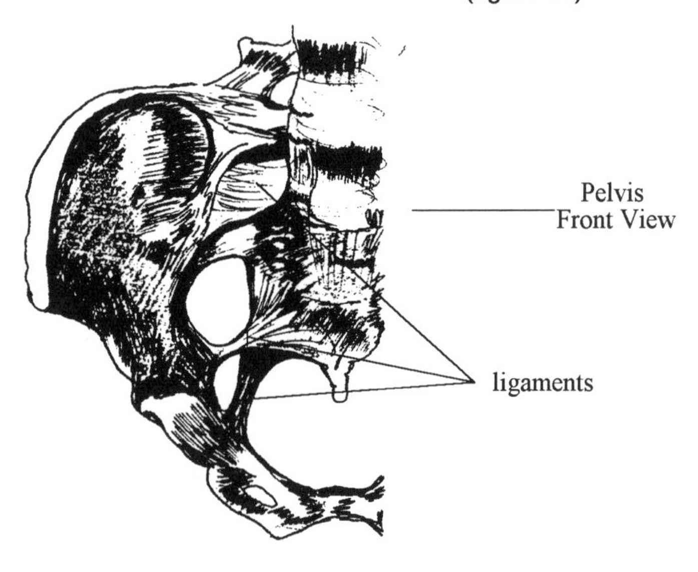

(figure 15)

Chapter Two
Function

The psoas is a multijoint muscle, attaching to six joints and passing over two. It can contract or shorten, as well as stretch or lengthen, in sections depending on the neuromuscular habits of a person and the work being required of it. The psoas is basically a **hip flexor**, although it functions by lengthening and falling back along the spine. It supports the free swing of the leg in walking, and plays an important role in transferring weight through the trunk into the legs and feet. More importantly, it is a **tensile structure;** a guide wire that stabilizes the spine. Picture a circus tent with its main pole and guide wires stabilizing the pole - a skin covering. The psoas muscle supports the spine as guide wires support a main tent pole.

As a **psoatic shelf**, the psoas muscle provides support for the organs and viscera. The health, length and vitality of the psoas muscle affects organ functioning. Whether or not there is room within the pelvic bowel for the organs to rest comfortably and function normally is determined by the length and tone of the psoas muscle.

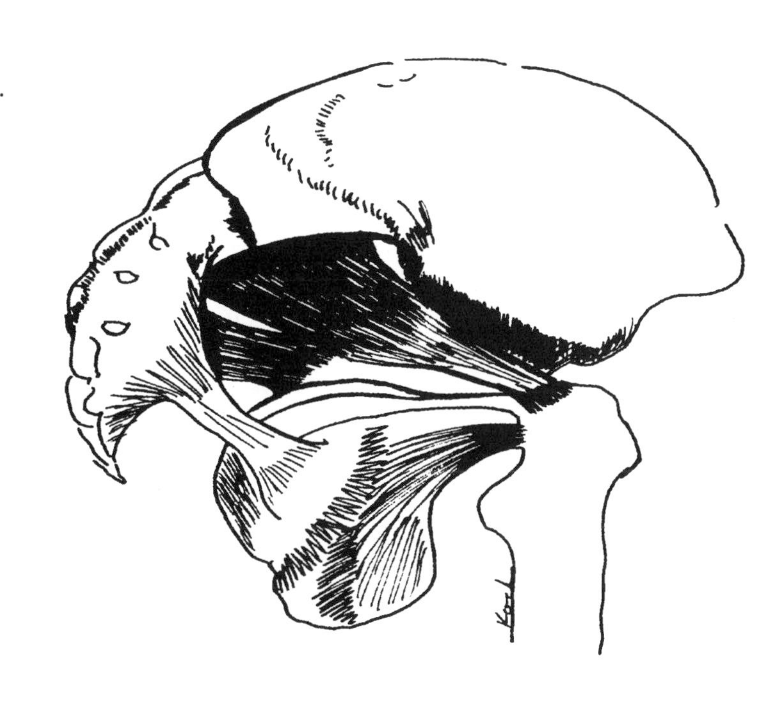

(figure 16)

The psoas muscle and obturators counterbalance each other. The obturators are a group of small muscles that attach within the pelvis (figure 16) They travel backward from the pubis, passing through the great sciatic notches, and insert into the femora just behind the iliopsoas. The psoas and obturators are opposing muscles; when in balance they create a pelvic base that is free-swinging, allowing freedom of movement to the legs. Ida Rolf in her book ***Rolfing - The Integration of the Human Structure*** explains the importance of the psoas in walking. *"Sturdy, balanced walking in which the leg is flexed through activation of the psoas, not of the rectus femoris thus invovles the entire body at its core level."* (11, pg 118) **Walking is a movement that is initiated not in the legs or leg muscles such as the rectus femoris muscle, but in the trunk of the body. The psoas muscle transmits or responds to the subtle shift of gravity through the trunk (the center of gravity) and the leg simply follows.**

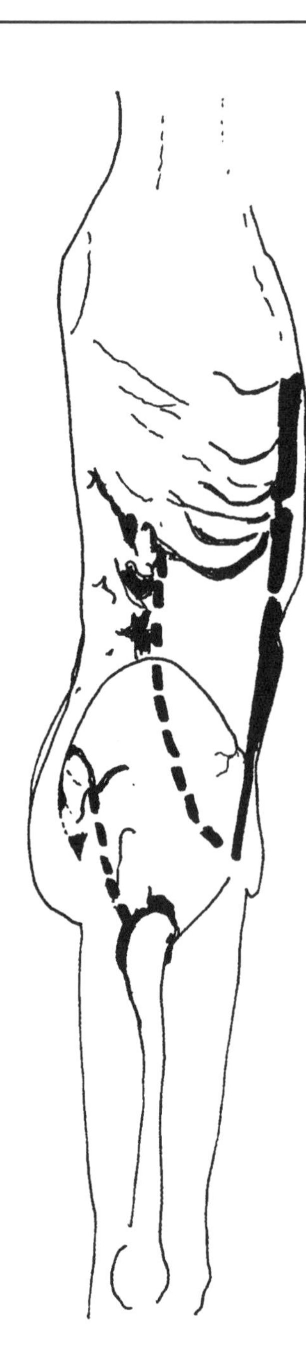

(figure 17)

The psoas and the erector spinea muscles have a reciprocal relationship. In the cervical spine the spinea muscles correspond to the lumbar psoas. Running along the posterior spine, the erector spinea muscles are often considered 'weak". However, when the psoas muscle is released and of a resting length not to obstruct skeletal balance, the erector spinea muscles begin to have a tone that supports the weight of the rib cage.

The psoas functions as a counterbalance to the rector abdominal muscles, by maintaining a front-back (anterior-posterior) relationship.

Function

(figure 18)

The appropriate balance between the rectors and the psoas helps to bring a sense of wholeness or functional relationship to the trunk. The lack of this relationship often leads to a chest out – belly in posture. Author Karlfried Graf Von Durheim in his book ***Hara*, *The Vital Centre of Man*** writes about the *"chest out-belly in posture and how it misses the natural structure of the human body".* He clearly suggests that when our center of gravity shifts upward towards the chest it forces us to *"swing between hypertension and slackness."* (4pg.17)

It is no surprise that, through sports and exercises, our culture concentrates so intensely and invests so much time and money in strengthening the abdominal muscles. (figure 18) We have lost contact with the deeper levels of our structure, and this concentration on the abdominals reflects our superficial attitude. Exercises like sit-ups and push-ups, not only weaken the psoas muscle, causing it to tense and shorten, but also provokes additional stress to the already strained back muscles, diaphragm, and viscera. This stress further decreases their efficiency and motility, while reducing the likelihood of our being able to sense the deeper, more quiet aspects of our being. The natural harmony and rhythm are lost, and the person, as reflected in the body, is fragmented.

Last but not in the least, the psoas functions as a **hydraulic pump**. Its movement stimulates and pushes fluids in and out of cells. In normal walking, the psoas muscle is activated, contracting and stretching with every step. Its' normal range of motion continually stimulates the viscera and massages the spinal column. This ability for the psoas muscle to move freely as a muscle rather than be used as a structural support, encourages a continuous, unobstructed flow of blood through the major arteries into the legs and feet. (figure 21) the psoas and the lumbar plexus, which enervates it, directly supply the energy needed to animate the legs, and plays an important role in activating both the anal and the sexual function.

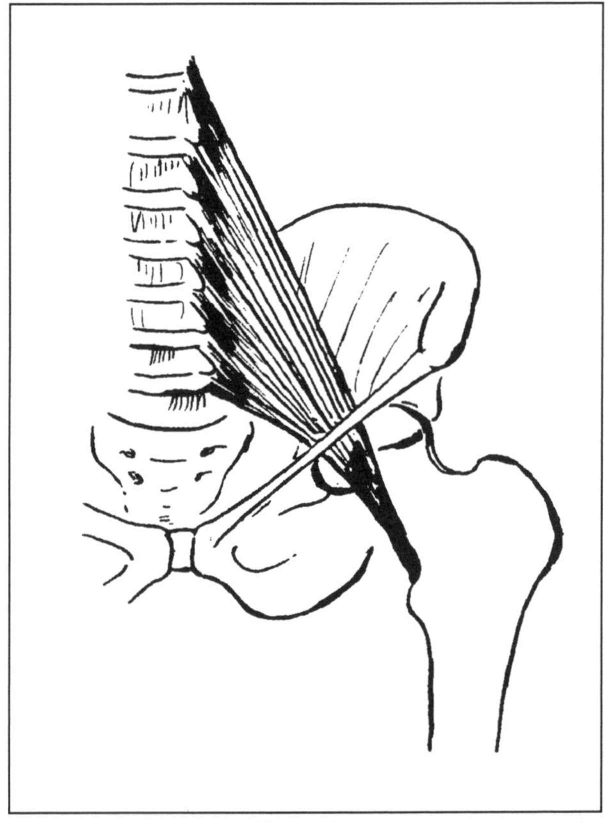

(figure 19)

Chapter Three
Effects

A few more basic concepts need to be clarified before all the varied effects of the psoas on the living organism can be understood.

To begin with, the main function of the bony skeleton is to support weight and to resist gravity. The prime function of the muscles is to move the bones. The muscles should not be used to support weight, as this disrupts their ability to move the bony levers freely, resulting in muscle fatigue. In time, the use of muscles to support weight can injure the bones even effecting their shape, and decrease blood circulation in the surrounding tissue, ligaments and joints.

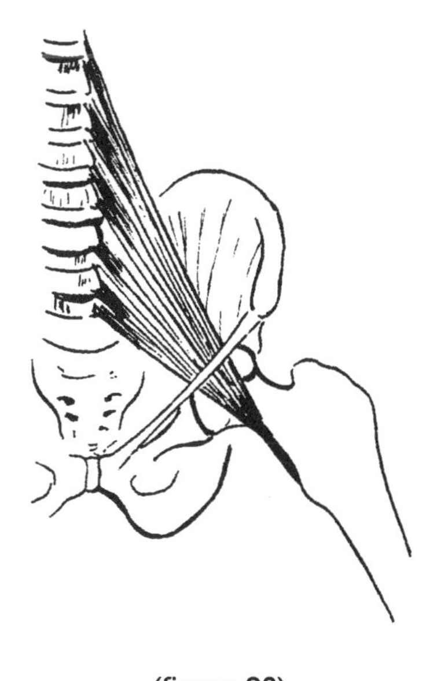

(figure 20)

It is the relationship of the bones to one another that allows the joints to transfer weight properly. Support is not something solid rather it is a matter of relationship both inside ourselves and to the outer world. Flexibility is experienced when the subtle energies, free to move without obstruction through the body, link the joints together like pearls on a string.

The psoas is a core muscle. Its' condition and resting length (that is its' theoretical condition when neither stretching or contracting), along with the forces of gravity (which may be seen and experienced as manifesting in spirals), directly influence the relationships of the bones to one another, and in turn influence freedom of movement in the joints. This influence begins when a child first encounters the force of gravity while sitting and standing. Whether or not the psoas is used to support weight, rather than the weight passing through the bones, will in part determine the postural stance of the person as he or she matures.

Effects

A shortened psoas, the result of continual contraction and miss-use, may be seen as a forward thrusting of the pelvis, creating rotational differences in the spine, pelvis and legs. These are seen as exaggerated curves in the lumbar, thoracic and cervical spine. These exaggerated curves are called lordosis - lumbar hyperextension and lumbar abduction to one side; kyphosis--increased thoracic curve; and scoliosis- lateral twisting curves in the thoracic and cervical spine.

In the pelvis the results of a shortened psoas include forward thrust, lateral tilt, and twisting. In the femurs, effects include differences in lateral and medial rotation. Thus the relationship between skeletal parts become fragmented, giving the body a disjointed, unconnected appearance.

The lateral differences in the psoas resting length will affect not only posture, but also the range of movement possible. If the psoas is more contracted on one side, it may cause that leg to shorten. This condition is often compensated for by putting a lift in one shoe. If instead the psoas muscle on the shorter side were allowed to release and lengthen, in many cases the condition would be remedied.

Why the psoas shortens more on one side than the other is an interesting question that leads into a new way of perceiving our bodies as the moving, changing organisms they are, rather than as static forms. The psoas may be shorter on one side as the result of accidental injury to that side of the body, resulting in exaggerated patterns of compensation for the injury. Or it may be due to the person's unique pattern of holding tension.

Whether it is movement within the body or the body moving through space, movement occurs in spirals. All living forms respond to invisible forces such as gravity, creating very specific patterns. One such pattern is the spiral. Look at the whorls of your fingertips, the shells found on the beach, the pine cones, plants and ferns of the forest and you will see the spiral pattern. Water flows down the drain in a spiral and the hair on our heads grow in swirls. If we use a microscope we see spirals in the helical structure of DNA. A telescope reveals the galaxies flowing in spirals.

Effects

The body is not static and it is not symmetrical. Although symmetrical movement is seen in very young children, as the child matures neuromuscularly, movement progresses into spiral and diagonal patterns. In her book, *Inherent Movement Patterns In Man,* Susan Campbell explains walking and running is a *symmetrical-to-diagonal* progression. As a child begins to bear weight and then move forward her first steps are *"taken with the arms raised in a protective symmetrical position and the legs advance in a primitive external rotation-abduction pattern.. The wide base of support in the early stages of walking demonstrates the unreliability of equilibrium reaction at this age." (2 pg. 55)* As the child matures the pattern moves towards an a-symmetrical, diagonal pattern. The wide base of support closes and *"the confident toddler demonstrates arm swings which are reciprocally patterned with leg movements, narrowed bases of support and leg patterns which are spiral and diagonal in directions."* (2 pg. 55)

Awareness of these diagonal spiral patterns can be used to develop better motor skills, whether the activity is a form of sports, dance or a daily job. Most lower back, neck and knee injuries can be considered the result of faulty posture, which causes structural points of stress that break down under additional stress.

When the resting length of the psoas is short, it may cause the pelvis to flex, limiting movement in the hip sockets by shortening the distance between the hip and thigh. When this joint is frozen the pelvis and leg move as a unit rather than as separate and coordinating parts. In turn the range of movement at the hip socket is limited. Instead of rotation occurring as the femur head moves within the pelvic socket, turning takes place through twisting at the knee and in the spine(L 4 and L 5). Whenever the excursion length of a muscle (the length of the muscle in movement) does not allow the skeleton to be placed or moved in such a way as to allow gravity to be used, it becomes impossible to perform certain movements and positions. This invariably results in substitution by other muscles, thus creating muscle strain, exhaustion and overall tension.

Effects

When the psoas is shortened or not properly engaged the ribcage will be thrust forward, encouraging thoracic rather than abdominal breathing. When pulled forward and down, the ribcage stops the diaphragm from descending fully; its' range of motion is lost. When the diaphragm can no longer fulfill its' function of stimulating and massaging the organs, nerves and blood, circulation may be impaired.

The major aorta, which is juxtaposed with the diaphragm and psoas, affects and is affected by the condition of the psoas, especially where they mutually pass over the hip socket.(figure 21)

A contracted psoas will shorten the trunk, affecting the structural position of the skeleton and lessening the internal space available for the organs and viscera. Food absorption and basic elimination is disturbed as the metabolic rate is affected by the lumbar plexus and its' autonomic neighbors. When the psoas muscle is not engaged properly, nutritional exchanges suffer.

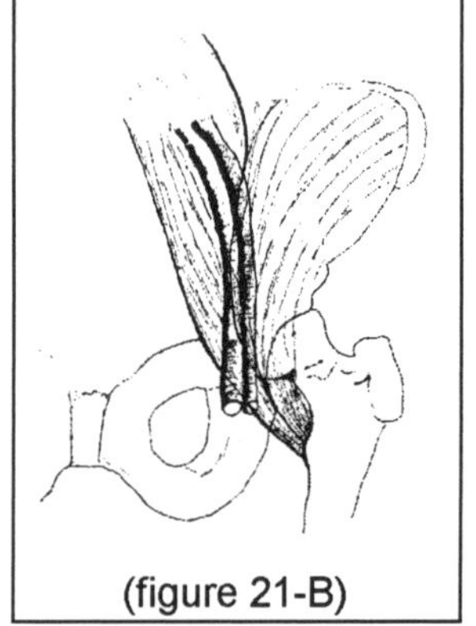

(figure 21-B)

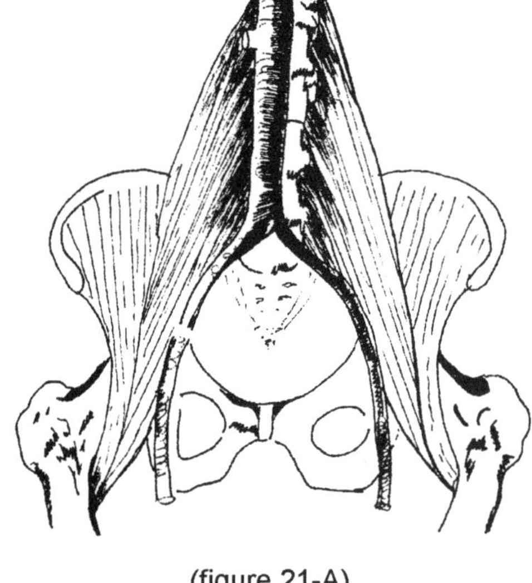

(figure 21-A)

Whether or not the autonomic ganglia and lumbar plexus are able to do their job effectively in turn affects the functioning of the adrenals and kidneys. The reproductive system too is affected, not only through the nerve supply but also through the positioning of the organs in the pelvis. For many women, menstrual cramps are not the result of cramping inside the uterus, but are caused by the psoas putting pressure on the organs.

Effects

When women are taught to release the psoas muscle, relief is very often experienced from cramping and lower back pain. (*see Women's Cycles Chapter)

Sciatic pain, a condition involving the major nerve running through the pelvis and down the back of the leg, (figure 22) may also be caused by, as well as relieved by, the health of the iliopsoas muscle group. Scoliosis, kyphosis and other structural problems can be seen and often remedied by working with the psoas muscle. An understanding of the influences of the psoas muscle on skeletal balance, muscular tone, and the health of the breath, nerve and viscera, builds the foundation for comprehending the indispensable role the psoas plays in having not only a healthy physical life but also a healthy emotional life.

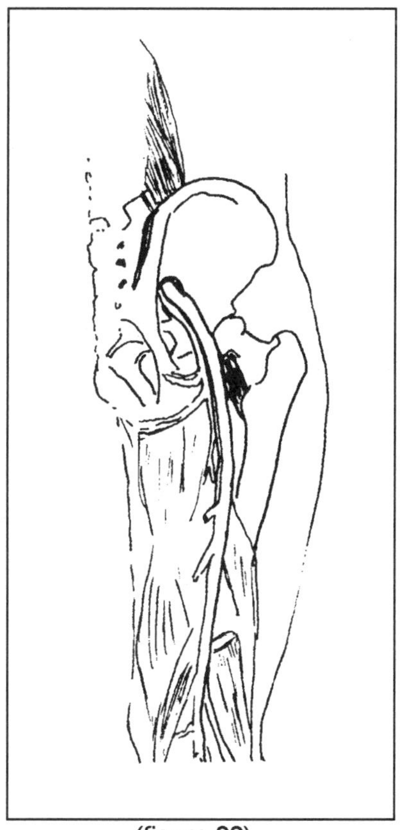

(figure 22)

Effects

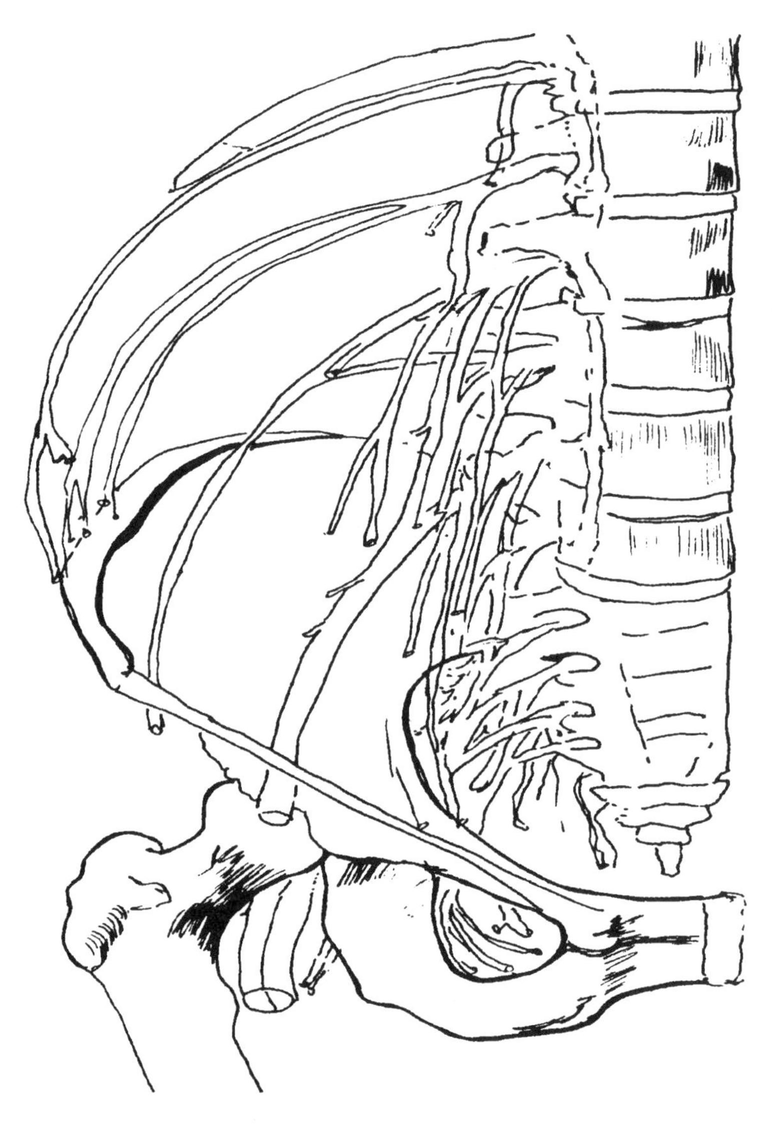

(figure 23)

Chapter Four
The Fear Reflex

Only one fear is considered instinctive and not dependent upon personal experience - the fear of falling. This reflex occurs through a complex set of nerve impulses, which basically command the flexor muscles to contract. Since the psoas muscle is a flexor, it is called into action as part of the fear reflex.

The sensation of falling is registered in the vestibular apparatus. Located in the ear, the vestibular mechanism is concerned with spatial orientation, motility, gravity and equilibrium. It is in very close proximity to the cochlea, which is part of the hearing mechanism. Any intense excitation occurring in either part will carry over and stimulate the other part. After the first three weeks of life, when the infant's hearing is more fully developed, loud noises will also stimulate the fear reflex. The reflex may be perceived as the organism closing in toward itself. Basically a fetal position, it is a form of defense, protection, and preparation.

The newborn startles when she hears a loud sound. Reaching out may be a primal response to grab hold of a supporting mother or tree branch to prevent her from falling. When the stimulus continues a second response may follow. The infant freezes - playing dead. In the book ***Body & Mature Behavior***, Moshe Feldenkrais explains "*---One would expect the first reaction to be such as to withdraw the animal from danger as quickly as possible. It is not so when the frightening stimulus is a violent contraction of all the flexor muscles, especially of the abdominal region, a halt in breathing soon followed by a whole series of vasomotor disturbances such as accelerated pulse, sweating, up to micturition and defecation...the initial flexor contraction---enables the animal to freeze and simulate death if the danger is too near." (6, pg. 83)*

Fear is the only instinct that has the power to immobilize. It is no surprise then that the psoas plays an important role in this reflex; the psoas is perhaps where the reflex gets its power to literally stop a person in their tracks.

The Fear Reflex

As part of the group of flexor muscles, the psoas contracts whenever the fear reflex fires. In so doing, it helps protect the person by bringing the extremities together, creating an enclosure that gives a sense of safety while protecting the soft, vulnerable parts: the genitals, vital organs, and the head and its contents - the eyes, ears, nose and mouth. The fear reflex also arches the spine, giving it more resilience and thus strengthening it against a possible blow.

Preparation to flee or fight involves the whole organism. The extensor muscles stretch and prepare for action, the pulse accelerates, and adrenaline is release into the blood stream. In normal conditioning, the reflex fulfills itself through the action of fleeing or fighting, and the body can return to a state of balanced relaxation. Noise pollution and diminished personal space produce and maintain a high level of stress. The natural functioning of the body is not oriented toward handling such constant forms of stress. Our ability to manage stress - to relax deeply and return to neutral functioning - depends greatly upon our conditioning and internal organization. It is within the framework of our conditioning, the maturation of the nervous system - that is, what we have learned - that the environmental stress factors must be met.

When the reflex fires repeatedly, with little recovery time and/or no ability to complete the cycle by fleeing or fighting, a conditioned response occurs that prepares the body for the next attack by maintaining a state of tension. Because the organism no longer has the opportunity to return to a normal state of functioning and nourish itself, the level of tension accumulates. This tension is experienced as anxiety.

As the excitation builds it radiates throughout the organs and nervous system. . Moshe Feldenkrais suggests *"Fear and anxiety are seen to be the sensation of impulses arriving at the central nervous system from the organs and viscera...there is ground for considering all emotions as excitation arising from the vegetable or autonomic nervous system and the organs, muscles, etc. that it enervates. The arrival of such impulses to*

---

the higher centers of the central nervous system is sensed as emotions." (6, pg.87)

When a child (and later the conditioned adult) is unable to defend him or herself by fighting or fleeing from a terrifying or hostile situation; and/or when the impressions received create an overwhelming sensation in the body, the individual may escape by "going into the imagination". It is as though the individual psychically removes him or herself from what is perceived as a life-threatening situation. By shutting out or cutting off the impressions, they "leave the situation".

Imagination is a useful tool that is inseparable from the ability to learn. When used intentionally it gives us the capacity to plan, to organize, to foresee our possible future and experiment with ideas and wishes we may not be ready to carry out physically. Imagination is the link between what we consider doing and what effects we think may or may not result.

Imagination is also a powerful form of escape. Although it may help to protect the child from a potentially dangerous situation, the price is high. Imagining the situation away does not physically satisfy or exhaust the fear reflex: instead, it disconnects the person from contact with the sensation, and the level of awareness diminishes. Eventually a deeper, more whole sensation of ourselves is lost, which, if felt, would move us to take action. The excitation however is still there, inundating the system so that the smallest reminder, such as an image, thought or sound, can elicit in us a sense of fear or nameless anxiety. As with all the flexor muscles, the psoas eventually shortens from misuse. The contracted psoas (and other flexors) by influencing the breath, establish a state or pattern of continual fear.

We can hold on to fear through our ability to image it, but what really gives life to the image or thought is the actual nerve excitation in the body. Fear is lodged in our bodies. It vibrates in the nervous system and is easily evoked. Although fear is often a subtle experience, we attempt to control this unpleasant feeling of anxiety by adding more muscular tension, resulting in layer upon layer of rigidity.

The patterns of holding tension appear not only in the physical body, but are expressed as well in the emotions and attitudes of the person. Muscular contractions are voluntary, (that is we can control whether or not we contract a muscle), and so stopping strong excitation, is a very empowering feeling. Rather than feel out of control or overwhelmed we can contract our muscles and diminsh or stop the intensity brought on by our sensations and feelings.

For every emotional state we eventually have a corresponding conditioned response. Catching the moment we begin to adapt a pattern of tension in any given situation is an enlightening opportunity for change. However in the young child muscular control is often the only means of feeling safe. By abating feelings and sensations of anxiety a child begins to get a handle on their emotions.. When habitually a child's feelings are not acknowledged and problem solving skills introduced, the child chooses whatever physical attitude works best to bring their feelings and sensations under control. Feldenkrais has noted that just such "*passive safety is brought about by flexor contraction and extensor inhibition.* (6 pg. 93)

Children learn to control their sensations and emotions by picking up verbal and nonverbal clues. Without being told how, we can stop crying almost on command. How is this possible, except by controlling and even immobilizing the body? The abdominals contract, the breath is held - basically, we freeze. The reciprocal relationship between the diaphragm and the psoas suggest that immobilizing one will by necessity influence the functioning of the other. Wilhelm Reich in his book ***The Function of the Orgasm*** **explains:** ***"The way in which our children accomplish this blocking off of sensation in the belly by way of respiration and abdominal pressure is typical and universal...The biological function of respiration is that of introducing oxygen and of eliminating carbon dioxide from the organism. The oxygen of the introduced air accomplishes the combustion of the*** *digested food in the organism. Chemically speaking combustion is everything that consists in the formation of compounds of body substance with oxygen. In combustion, energy is created. Without oxygen, there is no combustion, and consequently no production of energy. In the organism, energy is created through the combustion of food stuffs. In this process, heat and kinetic energy are created. Bio-electricity, also is created in this process of combustion. If respiration is reduced, less oxygen is introduced; only as much as is needed for the maintenance of life. If a smaller amount of energy is created in the organism, the vegetative impulses are less intense and consequently easier to master...reducing the production of energy in the organism, and thus, reducing the production of anxiety."* (10, pg. 276)

It is important to note here that working directly with the breath is not recommended except under the guidance of a qualified teacher. As breath is an essential function it is strongly suggested that one stay away from "breathing" exercises. There is no forcible way to change the condition of the breath, and most breathing exercises are not likely to result in change except change for the worse. It is the ability to let go of holding the breath - that is, getting out of the way of controlling the breath that allows proper functioning. Relaxing muscular tension by releasing the psoas helps to reestablish natural breathing.

"It is as absurd to learn to breath as it is to learn to make your blood circulate...Breathing needs not to be taught but liberated. It's imperfect because it's blocked. And it's blocked by causes that are foreign to the respiratory function. It's blocked by the shortening of the posterior muscles. The only way to treat a breathing inadequacy is, therefore, to make these muscles supple." (1, pg. 100)

For muscles to be supple the skeleton must fully bear the weight of the body, and the pelvis must be free to move. This allows the muscles of the pelvic floor, which are similar to the diaphragm, to be free to relax and "drop down". allowing sensation to circulate freely. The floor of the pelvis and genital/anal area are then experienced as open, vulnerable, and alive.

The sensation is not of localized tension but of deep, full relaxation. But if the psoas has become shortened it will eventually pull the skeleton out of the weight-bearing position, and will immobilize the pelvis. When this is combined with contracting the genitals and adductor muscles (muscles attaching from the pubis along the inside of the femur or thigh bone), the pelvis becomes static. Sensation in the genital, except through direct stimulation, is no longer experienced or is limited to a localized excitation. Sexual energy is then blocked by a frozen pelvis and a contracted psoas. To a body already blocked with fear, the powerful sensations of sexual energy will be experienced as overwhelming. The energy must circulate; if it is blocked from moving down through the genitals and legs, it will move upwards Dr. Wilhelm Reich explains in his book ***The Function of the Orgasm*** *"Sexuality and anxiety present two opposite directions of vegetative excitation. The same excitation which appears in the genital as pleasure, manifests itself as anxiety if it stimulates the cardiovascular system. That is in the latter case it appears as the exact opposite of pleasure."* (10,pg.110)

Fear and suppression of sexual energy affects the sense of well-being. The condition of the psoas is vital to experiencing full orgasmic potency. A full orgasm, where the whole body participates in an undulating reflex that takes over as voluntary control is given up, is possible only when the psoas is released. Once again it is the work of Moshe Feldenkrais clarifies the importance of releasing rigidity and muscular tension in the pelvis; *...by eliminating contraction and rigidity in the pelvic region, an obstacle interfering with reflex discharge of motor impulses, essential to normal orgasmic release of tension in the sexual act is removed; the way to complete maturity, sexual and otherwise, is cleared."* (6, pg. 15)

When the psoas releases spontaneously without the use of invasive or manipulative techniques, people often report elusive feelings of fear or a sense of general unrest. *"Often trembling in the lower half of the body takes place. The pelvis may begin to jerk spasmodically or rock and the legs vibrate. This is usually accompanied by anxiety or panic at first, marked by a holding of the breath...."*(7, pg. 66) As the psoas continues to release and the pelvis begins to extend, the "excitation" flowing chaotically at first gradually becomes "sensation" flowing smoothly through the whole body. The genital and anal areas are sensed as the pelvis relaxes and opens in depth and width. Feelings of vulnerability often accompany these sensations, and with them associations and attitudes about sex - in other words, the person begins to discover his or her conditioning as the connections to sensation are reestablished.

With the balancing of energy comes a balance of emotional states. From the point of view of the Oriental philosophy and art of medicine, emotions are qualities of organ/meridian functioning. Their manifestations depend upon whether or not energy is flowing without obstruction. Thus fear is observed as an imbalance of the *Qi* or *Chi* (life energy) which travels unceasingly through the meridians or pathways of the body. Author Felix Mann in his book ***Acupuncture; The Ancient Chinese Art of Healing and How it Works Scientifically*** explains that ;*"The ancient Chinese made not precise distinction between arteries, veins, lymphatic, nerves, tendons or meridians. They were concerned rather with a system of forces in the body, those forces which enable a man to move, to breathe, to digest his food, to think. As in other so-called primitive systems of medicine, like the Egyptians or the Aztec, the anatomical structures which make these physiological processes possible were not described in detail. They concentrated instead on this elaborate system of forces, whose interplay regulated all the functions of the body...Qi (life energy) is one of the fundamental concepts of Chinese thought. The manifestation of any invisible force, whether it be the growth of a plant, the movement of an arm or the deafening thunder of a storm, is called Qi".* (9, pg. 47)

The concept of energy is foreign to and rejected by many people raised in the West. And why shouldn't that be, when so often all a person is aware of in their own body is their tensions, as varied as those tensions may be? Many people are unaccustomed to sensing themselves; they may be unaware of the distinctions between sensation, emotions, and thought. It is as if they are locked in only one tiny room of a mansion, and have misplaced the keys.

The Fear Reflex

To actually perceive the fine energies flowing through the body would mean reestablishing those long-lost ties with sensation. It is a process that can't be done for us. Although others can guide us, we can't send our body in for a tune-up as we send in our automobile. For real contact to be established, the nervous system must mature. We have to pay the price of becoming aware of ourselves - as we are.

Where do the disturbances originate? Why is there social suppression of sexuality? Why are we taught not to touch ourselves, why do we stop sensing ourselves? Why do we need to control ourselves, nature - Mother Earth? What are we afraid of? It is the chicken-or-the-egg question; and yet all these questions are reflected in our bodies. Living somewhere above our waistlines, if not our shoulders, we speak of lower parts - down there! We seem to be trying to escape from the very link that nourishes us.

"...we can see how people offend against the harmonious relationship between heaven and earth either by straining and stretching upwards or sagging downwards...In both these cases the right centre of gravity – the one connecting the upper and lower is lacking. When it is present the energies pointing to heaven and those affirming the earth meet in harmony. What is above is supported from below. What is below has a natural upward tendency. The figure grows upwards from below as the crown of a tree rises from a vertical trunk, deeply rooted. Thus the right posture' expresses man's Yes to his bi-polar wholeness, his place between heaven and earth". **(4, pg. 81)**

The "right centre of gravity" of which Karlfried Graf Von Durkheim refers to in his book ***Hara, The Vital Centre of Man*** is experienced as a physical sensation and is perceived as an inward **attitude** known in Japanese as ***Hara.*** Hara literally means belly, and it's physically centered a little below the navel, the same area where the psoas resides. Balanced and harmonious movement can only occur when the psoas muscle is released and functionally engaged. Likewise, it is from the Hara that a balanced, harmonious movement arises, giving birth to a sense of integrated fluidity and wholeness.

Chapter Five
Childhood Conditioning

While we are in our mother's womb, a vital exchange takes place: nourishment is received through the umbilical cord directly into our center. Authors Ron Kurtz and Hector Prestera MD. in ***The Body Reveals, An Illustrated Guide to the Psychology of the Body*** write; *"If, in the uterus and early extra uterine environment, we are poorly nourished, we will find it emotionally difficult to receive any energies. Somewhere within us the experience of not receiving full warmth and love is imprinted. We will be untrusting, unable to open ourselves to the available nourishment around us."* (7, pg. 73)

Whether lack of nourishment is attributed to the feeling a woman has toward her unborn child or to the condition of her own impoverished pelvis, the psoas will have an influence on the condition of the woman during pregnancy and in turn on the unborn and new-born infant. (see Woman's Cycles Chapter).

It is easy to see the condition of the psoas in a pregnant woman by the location of her belly. Pregnancy offers an opportunity to release and lengthen the psoas. The additional weight naturally lengthens the psoas as it responds to the pelvic changes of pregnancy. However, this lengthening can occur only if the woman is structurally in a position in which her bones can support and balance her body weight. If not, she will find herself collapsing under the additional weight, and her baby will be thrust out in front of her. This is not to say that the belly shouldn't enlarge to accommodate the demands of the growing fetus. But if the psoas releases, much of the needed support will come from her pelvis so that the baby will be contained within her body, rather than sitting out in front of her like extra baggage.

At the moment of birth the infant's nervous system and brain, responding to life impressions, actually begin to grow and take form. It is the encounter with life that develops the nerve pathways. With maturation, the infant develops from a smell/touch-oriented being into a sight-oriented being. The vertical standing posture develops through a progression of stages where posture is coordinated through simple bodily reflexes until eventually higher coordinating centers regulate voluntary movement. This voluntary control of posture takes the child nearly three years to develop. The impressions of motor behavior perceived by the child in its first five years form the basis for all future movement.

Unlike an animal, whose nervous system and brain are more complete and defined at birth (a colt for example can stand and walk shortly after birth) the human new-born is dependent on adults until a much older age. Walking, for example does not take place immediately but evolves over a period of time. What this open system gives us is an incredible ability to adapt to varied life experiences and to develop in unique individual ways.

Inherent movement - the basic ways the human body moves - are inherited and are limited to a few simple patterns, which are recombined and reorganized to create more sophisticated patterns as the child matures. Each stage of motor development builds on the one before. The fundamental sequential stages leading to upright locomotion include: pivoting in the prone position, crawling in the prone position, creeping on the hands and knees, standing and balancing on two legs, standing and balancing on one leg, walking and running. At each stage a greater development of equilibrium is necessary to resist the force of gravity. The infant begins practicing resisting gravity immediately or as soon as she begins to lift or control her head. She does this in the prone position by extending the head and her trunk and by bilaterally flexing and extending her arms and legs. By stretching the extensors muscles the infant begins to gain the muscle strength necessary to resist gravitys' forces.

The body has the ability to align itself in space without our knowing how, or even being aware of just what is taking place. To understand that the body rights itself in space is to understand that the nervous system is equipped to do the job if we do not interfere. What we can determine is when we start and stop, and the direction, range, force and speed of a movement. Once decided, all we have to do is to "get out of the way" so to speak. What usually interferes is our conditioning. Our conditioning affects the maturation of the inherent patterns.

Through imitation and repetition the nervous system establishes habits of posture, attitudes, and movements that may or may not be appropriate to the situation. Moshe Feldenkrais in his book ***Body and Mature Behavior*** explains; *"In most of the acts taught to us, the insistence is on a procedure similar to that which the adult considers to be satisfactory...The child is urged in one way or another to do something to itself, and to become master of its vegetative system - often before he has even the rudiments of control of voluntary muscles...Some parents will pay more attention to this or that activity, some to another, depending on the age, the society and the knowledge prevailing at the time. It is thus a matter of pure chance, with a large bias towards the imitation of the adults concerned, as to which particular patterns of doing the child will strive for. Even in the best cases. it is most likely that he will adopt a particular manner of doing in order to satisfy or imitate some adult. Repetition...soon facilitates the flow of nervous impulses, as if the associated paths straighten, deepen and become preferred."* (6, pg. 151)

A small child learning to stand will use its psoas to help stabilize the growing bones. Working as a guide wire, the psoas helps to support the spine. However, as the weight passes naturally through the bones, the psoas gives up assisting the bones as a form of support. This occurs naturally only if the child's nervous system has the opportunity to mature, instead of the child being encouraged to sit up and walk before the action is initiated and carried out unaided. Holding a baby's hands above his or her head (with arms above shoulders) and "walking her" for example, does not quicken development but deters it. Each stage of development is a preparation and foundation for the next. Crawling as another example, is a means of developing and maturing the hip sockets. Playpens and confining apparatus limit movement. And yet it is through movement that the nervous system and motor skills develop. The stage of crawling develops and refines the hip sockets. Sufficient extension of the hip joints is possible only when there is an adequate period of crawling. When a baby is not free to move, that is restrained by various apparatus for long periods of time (such as the seats used as both day containers, car seats and portable beds) she looses the opportunity to develop important motor skills. Prolonged standing before a baby's bones are weight-bearing conditions her to associate standing with an improper muscular pattern. In pre-mature standing the spine has not formed its lumbar curve and the muscles of the lower abdominal wall are not developed.

If a baby is *walked* (i.e. held by her arms before she is able to stand upright without support) her sense of stability and safety become precarious or lost and reliance on the psoas increases. On the other hand, a child who is not pushed or encouraged to walk before he voluntarily initiates the full movement will do so at a "later" time - approximately 16 to 18 months. When the child does stand and walk, it is with skill, balance, agility, and confidence. When the child falls, it is more like sitting down. This level of coordination is then reflected in all motor skills.

Walking is very much an act of falling and catching. At the moment when the body begins to fall forward, the psoas is released and the righting reflex fires, antigravity muscles (mostly extensors) activate, and a moment of catching occurs. The body walks from its center, rather than being pushed and pulled along by the legs and arms.

For the sensitive nerves located in the heel and ball of the foot to fire, mobility in the foot is necessary. Shoes therefore can play a very important part in learning to walk freely. Hard-soled shoes and stiff high tops confine the foot and ankle limiting movement throughout the body. Learning to walk (as with all the range of movement skills) needs not to be impeded by artificial limitations created by our environment. Gaining a natural stability in the beginning of life is much easier than having to learn new movement patterns later in life.

Chapter Six
Releasing The Psoas Muscle

The psoas muscle can be activated again as a functional part of the organism. The psoas needs to lengthen before appropriate functioning can take place. To release and lengthen the psoas involves time. A good place to begin is the ***"constructive rest position"***. Lulu Sweigard (see reference). coined the name for this spontaneous position in the 1930's. The position allows the psoas to release naturally as the person relaxes.

Choose a place to work that is quiet, and where no one will be stepping over or around you. Choosing a safe, quiet place to work, allows you to relax and open to your sensations.

To Begin: Lie down on a padded floor, on your back, with your knees bent at a 45 degree angle,and your feet on the floor. Separate your feet and your knees the width of your hip sockets (remember they are located in the front of your pelvis). Place a towel no thicker than 1 1/2 inches under the back of your head. Fold the towel neatly so that it is flat and place it under the upper half of the skull, *not under your neck*. The function of the towel is to support the cervical spine so that it is at the same level all along **the back. The head should not tilt forward toward the chest or backward toward the floor. Allow your arms to rest at the sides of your body, on your pelvis, or cross your arms and rest them over your chest. Allow your eyes to remain open but resting deep and soft within the eye sockets.**

When you first lie down, if you notice that your back is arched, leave it; do not try to force it to the floor. **KEY:** ***use no force.*** Instead of "trying" to get physically comfortable, allow your thoughts to quiet, and bring your attention to sensing your body, especially the inside of your pelvic region. As you lie there, the lower spine will begin to have weight and will eventually release and lengthen along the floor all by itself.without you forcing it to happen. This is a result of the psoas muscle releasing. In this position gravity influences the skeleton by releasing the psoas. Force is neither needed nor helpful!

When you are ready to leave the position, roll over and rest for a moment. Get up slowly. ***Don't pull yourself up out of the position with***

your neck muscles. Use your hands and legs to bring you out of the position. Once on your feet, take the time to observe any differences you sense. How do you sense when standing up?

Releasing the psoas muscle will help relieve many discomforts while re-establishing appropriate skeletal balance, muscular and visceral functioning Remember the intricate connections among the various functions means that it is more complicated than just releasing one muscle. And yet, releasing and lengthening the psoas is the beginning of changing the whole person. To change ones' posture involves more than merely stretching the psoas muscle. For there to be harmonious long-term results, the work needs to be in the context of your life and your habits of living. Change comes from self-understanding.

The *constructive rest position* is a good position to use when teaching clients and students. It is a subtle and non invasive method of assisting the person in releasing their psoas. The awareness methods cannot, unfortunately, be understood through reading but must be experienced first-hand. (see Resources - Training Workshops)

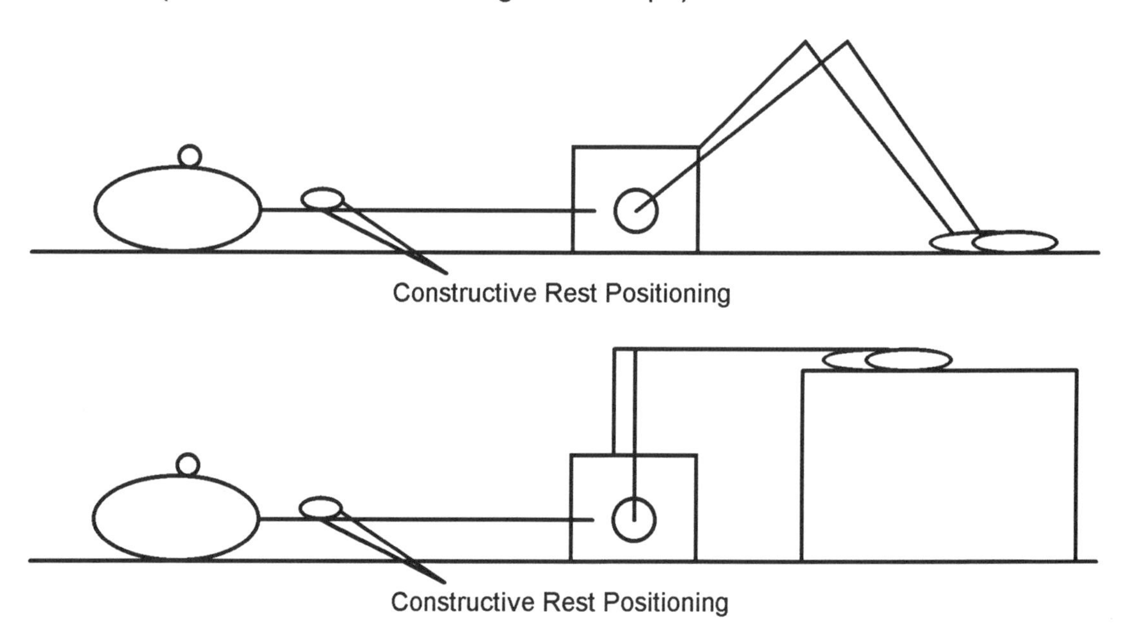

Constructive Rest Positioning

Constructive Rest Positioning

Releasing The Psoas Muscle

Lying down in the *constructive rest position* for ten to twenty minutes every day revitalizes the body and prepares it for the day's activities. The best times to work are in the morning and after the day's work, (before dinner). Lying on the floor frees the central nervous system from much of the stimuli that evokes habitual response patterns to gravity. In the beginning, subtler sensations will be experienced in the rest position than is possible while standing.

Most of the discomfort you experience in the *constructive rest position*, whether physical, (such as achyness or tension) or emotional, (such as feeling angry, sad, frustrated or scared) arise from your conditioned muscular patterns and not from the immediate situation. As the psoas releases, the sensations you experience may cause you to feel vulnerable. Instead of changing or re-arranging your position, try to become more quiet, and simply follow the sensations as they circulate freely through your body.

Notice the stream of images, thoughts and emotions that move through you as the sensations appear. There is no need to change anything, or for your attention to be absorbed by these associations. Instead, return to the qualities of the sensation - not only those inside your body but also the sensations experienced by incoming impressions. Sensations such as the air currents on your skin, the warmth from the sunlight in the room, the odors and fragrances in the air, the texture, temperature and pressure of the floor and mat, and the sounds all around you.

Match the impressions coming into your body with the quality of sensation and feeling inside of your body. A balance will begin to occur between what is experienced internally and what is being received in the moment. By balancing the intensity of your internal world with the impressions (new food) coming into you from outside of your body and observing the moment that the two interface, is a process of awakening Being in the moment forges new nerve pathways by letting go of old conditioning.

Releasing The Psoas Muscle

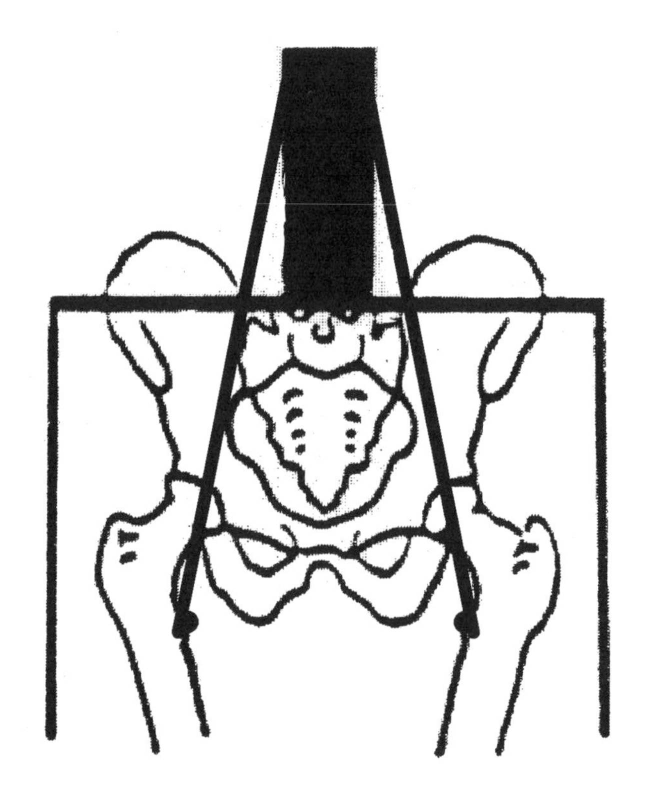

A Stable Pelvis Frees The Psoas Muscle To Be A Guide Wire

Releasing The Psoas Muscle

Working With Your Psoas Muscle

- Always begin with the **Constructive Rest Position**!

- First notice your individual torquing pattern. What parts of your body have weight? What parts sense as lifted or held off the ground? Begin by simply following where your sensations lead you without trying to change anything.

- As you stay in the ***constructive rest position*** notice your feelings, what is the emotional quality of your pelvis. How does it feel to you? If you were to draw a picture of your pelvis what images come to mind? What colors would you use?

- **As you rest in the *constructive rest position*, your pelvis will begin to move into extension, the back of the pelvis will widen and flatten as your iliopsoas muscle releases. The spine begins to elongate and the shoulder girdle widens.**

- As the psoas releases there is more awareness and openness in your hip sockets.

- Bring awareness to the front of the hip sockets on the top of your pelvis (approximately on each side of pubis bone). Place your fingers on your hip sockets, they will help to attract your attention to the area.

- As the psoas releases, and frees the hip sockets notice how weight begins to pass through your legs into your feet.

- Allow your eyes to be stay softly open (do not look out to the ceiling, but let the light in the room come to you).

- Match your sensations, feelings and thoughts with incoming sensory impressions (from light, sound, touch and smell).

Working with Your Psoas

Releasing The Psoas

Once you have rested in the *constructive rest position* you are ready to loosen the hip joint and work with differentiating the leg from the trunk of the body. The separation takes place at the hip joint. Begin by lifting one leg to help stabilize the pelvis. Do not bring the leg to the chest, just hold it at arms length. You will be working with the opposite leg (the leg and foot that are still in contact with the floor). Begin slowly sliding the foot down towards the floor. Keeping your attention on the hip socket, **move only the leg not the pelvis. KEY:** When doing this exercise properly, there are no sensations in the lower back. If your back arches you are moving the pelvis with the leg. Instead keep your awareness at the hip sockets. Releasing the psoas muscle will allow the leg to move without the trunk. It is important to stop when you reach your excursion length of the psoas. **Use No Force** The purpose of the exercise is not to extend the leg fully, but to mature the hip socket by learning to voluntarily release your psoas muscle.

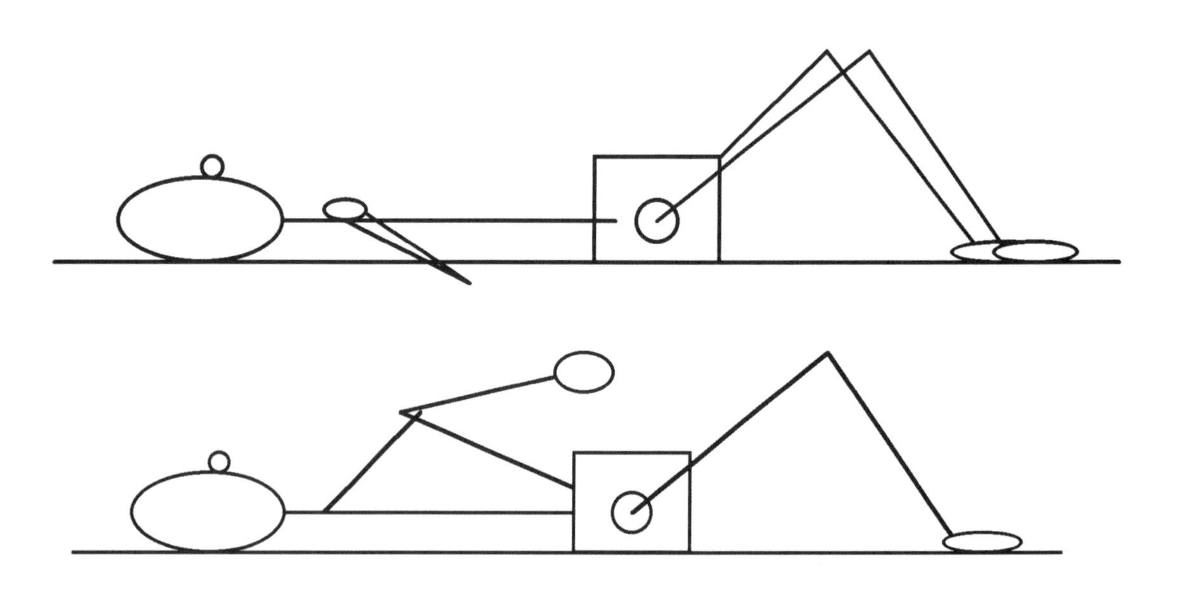

Releasing The Psoas Muscle

DO: Lower the leg slowly, keeping the pelvis as part of the trunk and only moving the leg at the hip socket.

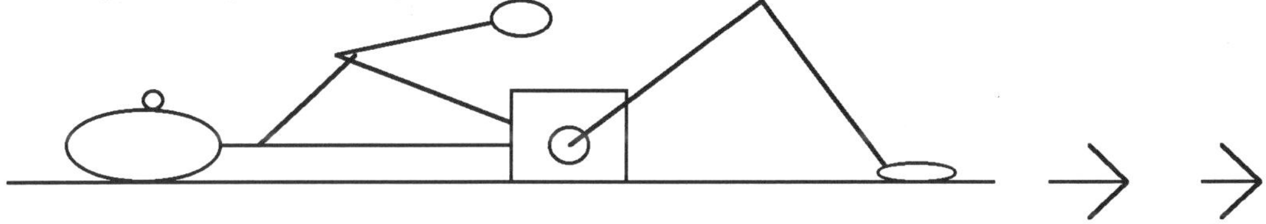

DO: Extend the leg only as far as it will go without bringing the pelvis.

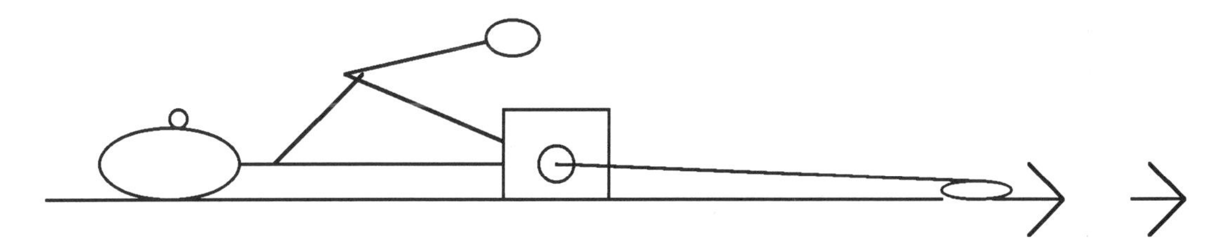

DO NOT: Move your pelvis with the leg, instead keep your awareness on the hip socket releasing.

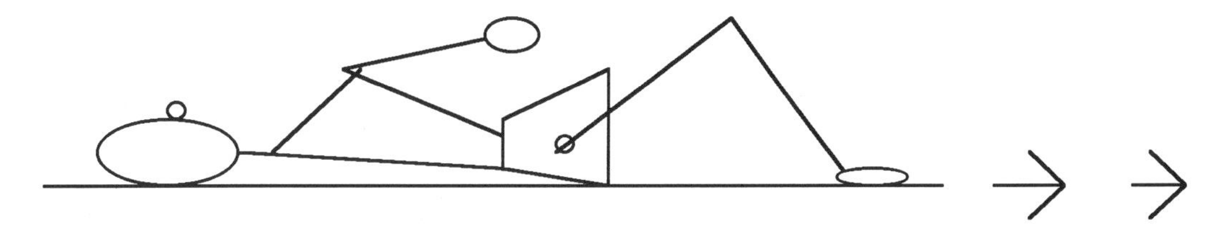

DO NOT: keep extending your leg if your pelvis is moving. Stop and bring more awareness to your hip socket.

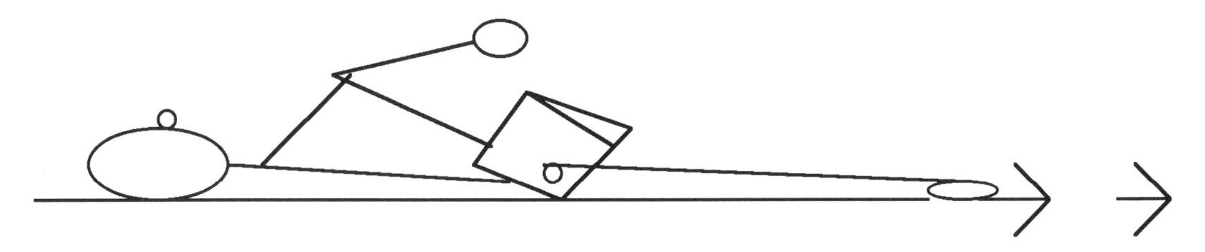

Releasing The Psoas Muscle

Releasing the Arms

Working with the arms involves freeing them from the trunk of the body. The psoas muscle needs to stay released. Begin by focusing your awareness into the hip sockets. Using no force, keep the pelvis and feet grounded by sensing your weight passing through the pelvis, legs and feet. Lift the arms straight with palms facing each other. Sense the weight of each arm as it releases at the shoulder socket. Begin slowly letting the arms move back towards the floor, over your head. Keep arms the distance apart of the shoulder sockets. If your weight stays in your pelvis, the arms begin to work like a lever; their weight stretches the upper psoas. You may be able to reach the floor or you may not. Do not push the arms to the floor by arching your lower back. Instead, keep your awareness focused in the front of your hip socket, letting go of any muscular holding. Return to the *constructive rest position*, by keeping your weight centered in your pelvis and feet. Pushing into the ground with your feet helps you lift the arms without pulling on the lower back or arching the spine. **KEY:** Allow the head to respond to the arms. The eyes and head follow the movement of the arms. **DO NOT lock the head**, let it move.

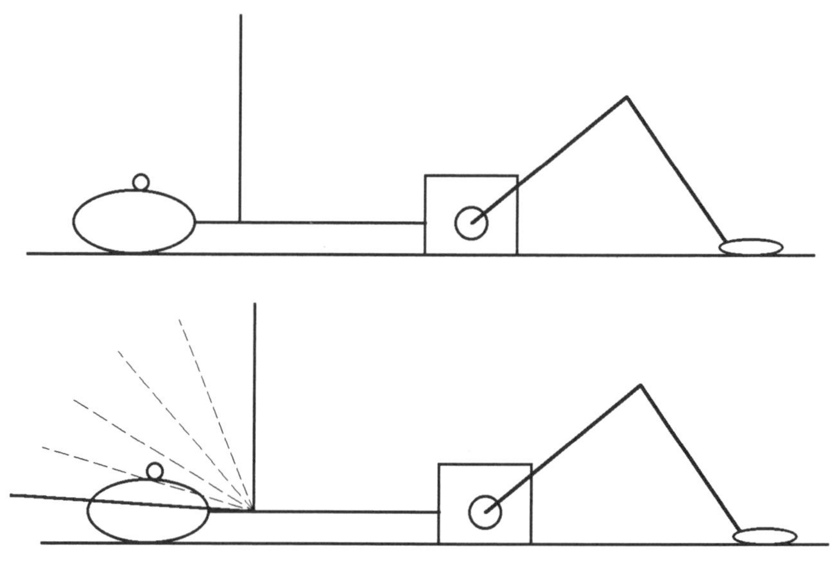

Releasing The Psoas Muscle

Toning The Psoas

KEY: Always keep your pelvis stable. Unaligned bones compromise joints and stretch ligaments. There should be no torquing through the trunk. Skeletal positioning must be maintained for proper muscles to be engaged. A released psoas is essential for toning the psoas and the surrounding muscles. If the psoas is contracted or tight, you will compensate and use the wrong muscle group to perform the movement. Many people have fatigued psoas muscles not weak ones. The exhausted muscle is thought to be weak because it is unresponsive. **KEY:** Resistance not force promotes strength.

Begin in the *constructive rest position,* and release your psoas muscle. When you can fully extend the leg without involving the pelvis you are ready to tone the psoas and leg muscles. Releasing at the hip socket, lift the extended leg the height of your hip socket (approximately 6 inches) and at the same time sense the weight pushing through the bent leg and foot that is in contact with the floor. The lift comes from pushing down with the opposite foot . Now slowly move the extended leg up and down, now sideways, back and forth and finally on the diagonal. **KEY:** When lifting your leg, think of the psoas muscle falling back along the spine and scooping the leg off the floor.

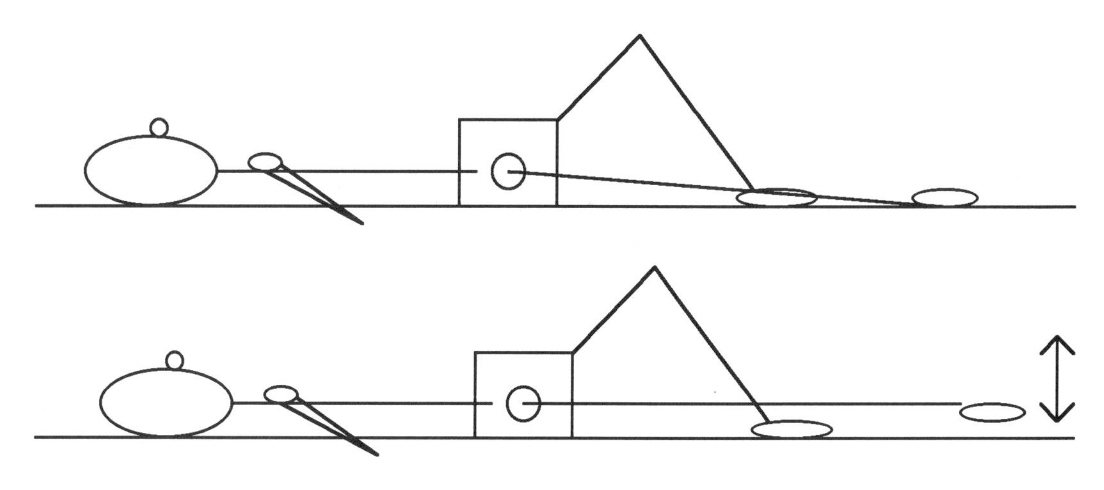

Releasing The Psoas Muscle

DO: Begin by distributing the weight through all four contact points.

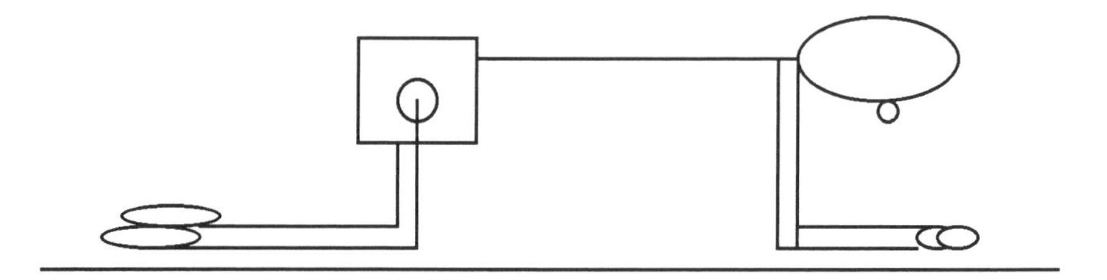

DO: Lengthen and extend one leg by releasing through the hip socket

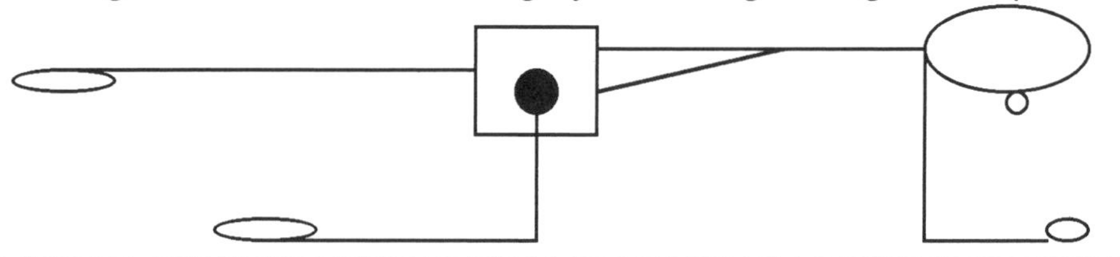

DO: Keep the pelvis stable and move the extended leg side to side, and up and down

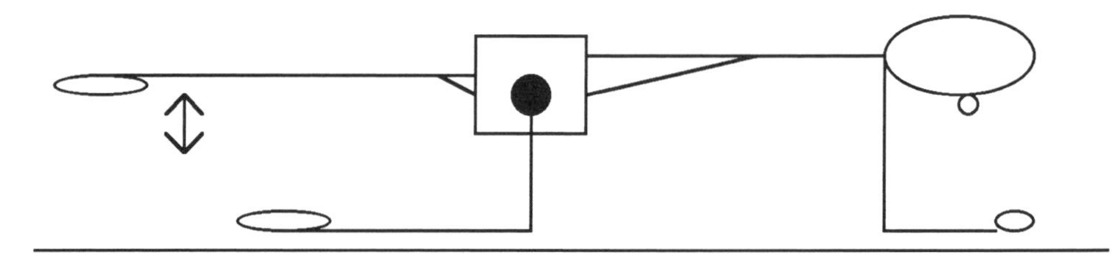

DO NOT: let the lower back and pelvis tip to move the extended leg. Keep the pelvis straight and release from the hip socket.

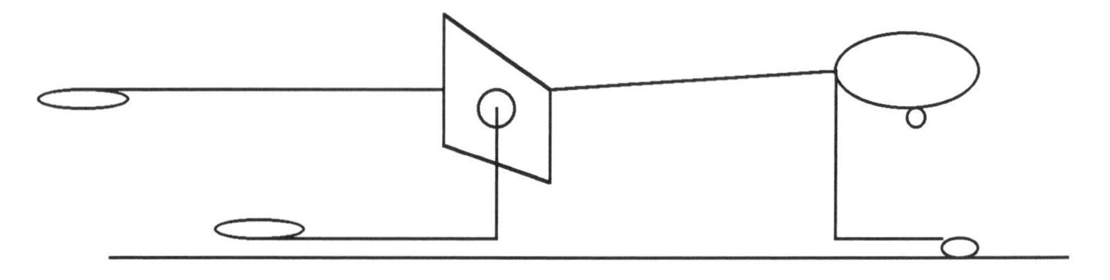

Releasing The Psoas Muscle

Lengthening The Psoas

To lengthen the psoas muscle you must release before stretching it. Releasing first allows the bones to be in a weight bearing position. Aligned joints are free to move. Lengthening the psoas frees the leg from the trunk increasing range of motion.

KEY: Keep your awareness in the hip sockets. Never strain or tense the lower back. The pelvis must be stable and the trunk unified. A unified trunk frees the psoas muscle giving it integrity to work as a whole muscle rather than in segments.

DO: Begin in the constructive rest position

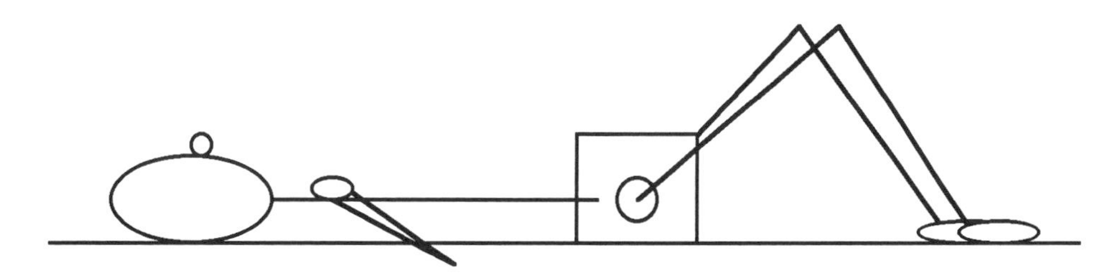

DO: Release through the hip socket and bring one leg to your chest

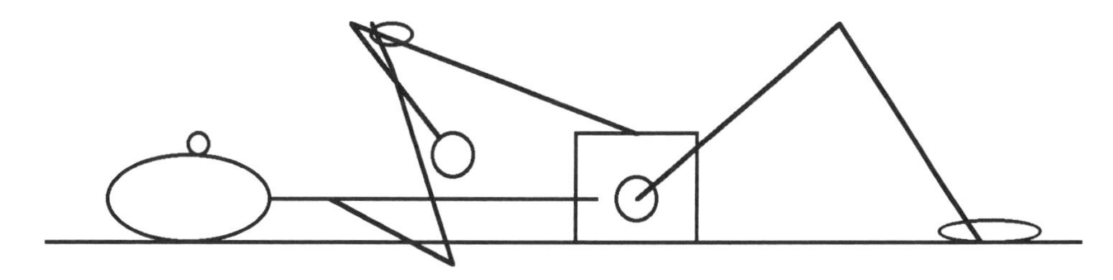

Releasing The Psoas Muscle

DO: Press the leg to the chest and then resist the position by pushing against the arm that hold the leg in place. Slowly slide the opposite leg down. Remember to keep releasing through both hip sockets. The psoas will stretch on the side that the leg is extending. It can stretch only if the pelvis does not move with the leg.

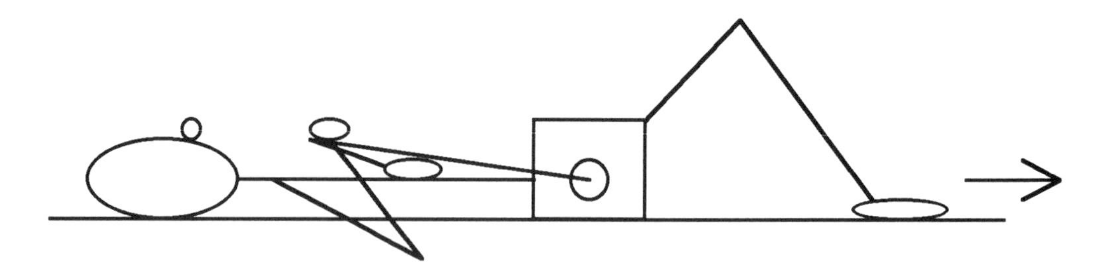

DO: Allow the leg to fully extended, keep holding yet resisting with the opposite side.

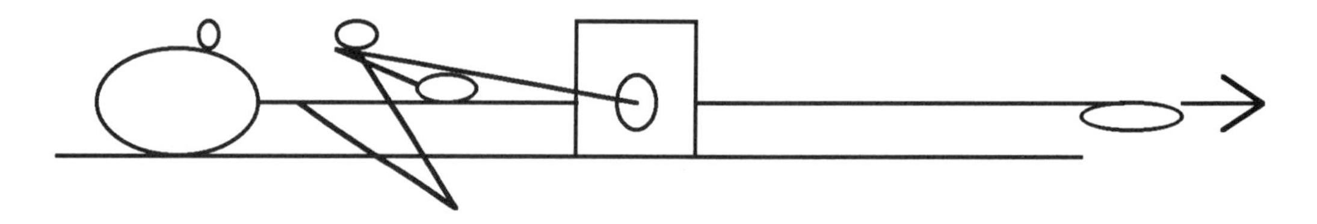

DO: To return to the constructive rest position first drop the flexed legs' foot to the floor

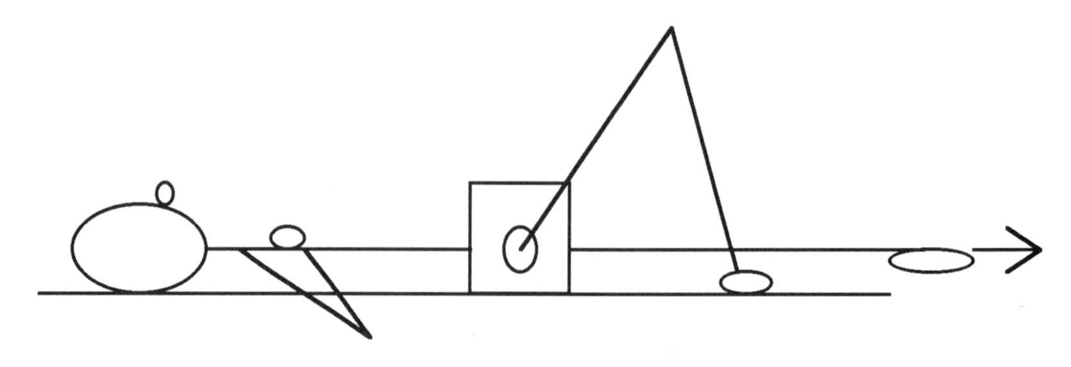

Chapter
Women's Cycles

Menstruation

The experience of menstrual cramping has been successfully relieved by releasing the psoas muscle. Women who are taking drugs for their severe cramps have found that releasing their psoas helped them to be pain and drug free. What is the relationship between the psoas muscle and cramping? Rather than the cramps occurring from within the uterus it is often a contracted psoas muscle that presses on the reproductive organs; constricting blood circulation and impeding, possibly irritating the nerves that enervate both the muscle and the specific organs. Because the psoas muscle is part of the fear reflex, fears associated with menstruating, reproduction and sexuality need to be understood and addressed.

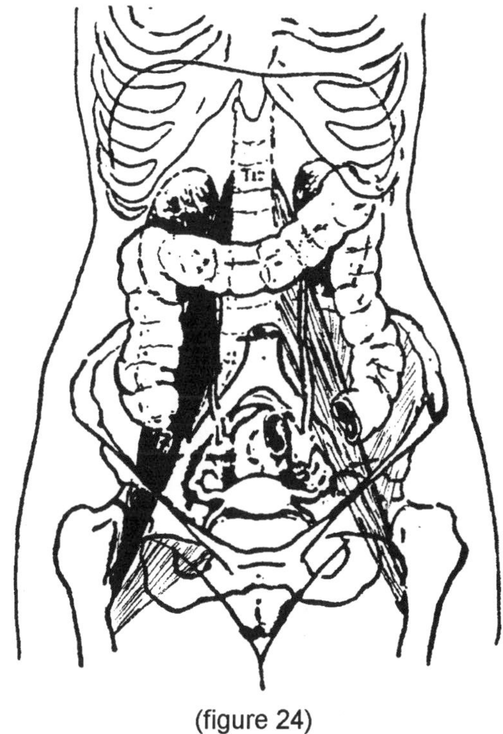

(figure 24)

A young girl begins her cycle of bleeding between the youthful age of 9 and 16. Insecurity, shame and embarrassment are often associated with menstruating. The feeling of one's body being out of control can feel very scary. Our culture has no rituals for assimilating, let alone acknowledging or celebrating the profound changes in a young girls life when her body begins to bleed. Traditional, religious and sexual taboos add to what is already an extraordinary experience.

Women's Cycles

She who bleeds but does not die -embracing this awesome experience, takes letting go of our childhood innocence and a willingness to emerge into the world as fertile women.

Fear is always sensed through the psoas muscle and so it is through sensing the psoas that we can release old fears.

Pregnancy and Birth

Pregnancy offers a natural opportunity to lengthen and tonify the psoas muscle. Due to the extra weight and increased awareness in the center of her body, a pregnant woman's attention naturally goes into the *Hara.* She experiences a *right center of gravity*; a feeling of centeredness which is truly an expansive and deepening sense of herself and her innate power.

If her psoas muscle is short and constricted, it reduces the internal space available for organs and viscera as well as the womb and growing fetus. When we feel afraid, our psoas muscle contracts, limiting space and volume in the trunk, and sending signals of stress throughout our body.

The following women discovered the powerful influential part their psoas muscle plays in the birth process.

- One month "overdue", a pregnant neighbor is scheduled for an induction. She is experienced with birth, knows that her babies take an extra month to grow and is aware that she is being pressured by medical guidelines to give birth. She is knowledgeable about how an induction will affect her labor negatively. She feels confused and pressured. Her home life is busy, with guests popping in from out of town. Her husband is supportive but preoccupied with work. She is feeling unable to nest and center on the coming birth. Together we release her psoas muscle and we both sense the baby's head move low within her pelvis. She feels calmer and her awareness focuses on the eminent birth. She cancels her induction and within 24 hours goes into spontaneous labor.

- A soon to be first time mother is referred to me by her obstetrician. He is concerned that her daily telephone calls reflect her fears and inability to face the challenge of birth and mothering. He hopes the session will help her relax. While releasing her psoas we seek to recognize the *energy of birth* and help her focus her attention on this energy flowing through her pelvis and down her legs, through her feet into the ground. Her new found sensations do awaken and deepen her trust in her body's ability to give birth. She leaves my office with an inner confidence and a physical stability. She has a long, hard labor and a smooth wonderful birth. She feels empowered as a woman and elated with herself and her new role as mother.

- A potential VBAC (Vaginal Birth After Cesarean) mom enters my office for her first visit. She is in light labor and is feeling ambivalent about the birth. Her visit is a "last ditch" effort to have a successful VBAC. She tells me she wants a vaginal birth but fears the unknown of labor. She already knows what to expect from a repeat cesarean. She walks around my office during mild contractions, breathing lightly. Her walk is ungrounded, her contractions offer no relief, her body lacks a focus. Her first baby never "dropped", meaning it never engaged in the pelvis ready to travel the birth canal. We work together, releasing her psoas. She confronts her resistance. Unacknowledged feelings begin to surface. She realizes she wants to birth on her own, surrounded and assisted only by other women. She does not want her husband present as she feels that she "gives her power away" in his presence. Her feelings of inhibition prevent her from doing what feels necessary to fully let go. It is an eye-opening experience and yet having no time to really assimilate her new found feelings her attempt at a VBAC fails and she has another cesarean.

- A very confident, older, woman pregnant with her first child is referred to

---

me for lower back discomfort. She has a history of back problems and during her pregnancy is experiencing sciatic nerve pain. She is articulate and clear as to what she wants from her health care providers and her birth experience. As we release her psoas, she begins to soften. Being in control gives way to trusting her body. Her center of gravity shifts as she experiences a deeper place of balance, and she experiences support from within. Her back pain is released and her sciatic discomfort is relieved.

In very practical terms, releasing the psoas means the pregnant woman offers her growing baby more room. Her pelvis becomes a bowl containing all of life. Rather than carry her baby out in front of her, feeling overwhelmed and off balance, she can draw inward strength and satisfaction from her growing belly.

When labor begins a released psoas encourages the downward flow of energy and assists her letting go of the baby. In a physically and emotionally trusting place, this can facilitate a smooth labor and birth. If there is disharmony, internal sensations and her feelings of fear will signal her to ask for the necessary changes.

FIRST TRIMESTER: Keeping the psoas muscle released through the of pregnancy can center a woman's awareness internally, releasing initial fears, helping relieve nausea by creating more volume throughout the abdomen. Keeping calm rather than ready to *flee or fight* helps to smooth hormonal changes. Releasing the psoas increases blood flow essential to the growing fetus, stimulating circulation and digestion so important for nourishing both mother and baby.

SECOND TRIMESTER: Releasing the psoas muscle creates more room for the growing fetus, stimulating organ functioning and helping accommodate the shifting support of the body. Releasing and tonifying the psoas relieves minor aches and pains and contributes to a sense of well being

Women's Cycles

THIRD TRIMESTER: Releasing the psoas and keeping it tonified relieves pressure from the discomfort of an ever growing belly. It strengthens the psoas as a shelf which supports the organs and increases over all energy. A released psoas muscle allows the hip sockets freedom to move and opens the area for transferring body weight through the legs. Learning to release the psoas prepares a woman for birth. By releasing fears and pent up emotions a woman attunes herself to *birth energy* (the energy or power within her that will *see her through* the birth), and clears spiritual and emotional channels for birth. Releasing the psoas muscle takes a woman out of her thinking mind and draws her attention into her belly, her *Hara,* right to her gut feelings, stimulating intuitive awareness and deepening her insights.

LABOR: Learning to release the psoas muscle enhances a woman's birth experience, shortens labor and in many cases can be an alternative to chemically and intrusive methods of induction. It should certainly be the first step before an induction is considered. A released iliopsoas assists the pelvic bones to open, facilitating the baby's movement through the birth canal. By not hindering the pelvic ligament's natural ability to softened hormonally for birth, and by helping to widen the pelvic bowl, the iliopsoas plays its part in gently letting the bones part for birth. Most importantly, I believe a woman who is really in touch with her center(through a released psoas muscle) will intuitively choose the very best place for the birth of her baby. Too often the choice of birthing location is made from fear based ideas of what a safe birthing place needs to be, rather than from an instinctively guided choice. When a women chooses from her gut feeling, her choice empowers and strengthens her for the incredible journey through birth into motherhood.

(See Chapter Releasing The Psoas)

Breastfeeding Although there are no obvious connection between breastfeeding and the psoas muscle, releasing your psoas will enhance your breastfeeding experience. Just as with birth, intuition, is a vital tool for breastfeeding. Intuition is your mothering guide helping you respond to your newborn and later your toddler or child's needs.

Rather than follow the whim of the culture, tapping into your deep roots will help you discover the joy and love that flows through you as a nursing mother. The more relaxed you are the more successful. The more centered, the less likely other peoples' opinions or your own doubts will stop your milk from flowing. As there is nothing short of breast surgery that prevents a women from breastfeeding, it is essentially an issue of feeling emotionally secure in oneself as a women and feeling supported by 'the loved ones around you. Body knowledge knows how to breastfeed. However at times, It can feel like a lost art. Finding other women who are experienced breastfeeding mothers is both beneficial for mother and baby.

Breastfeeding is a gift of nourishment and love that relaxes the baby's psoas muscle as well as the mothers. It fosters a deeply felt satisfaction and encourages a profound energetic bonding relationship.

Menopause A great time of change, menopauses' hormonal shifts demands that our attention be given to all aspects of our lives. As the psoas muscle helps stabilize the body, it is often called upon during this time of metamorphous. Organs adjusting to their new rhythms send messages, via nerve impulses, to the brain. In response, the vertebrae torque, (via the psoas muscle), in a two way conversation. A dance takes place that can either enhance ones' sense of oneself or create more feelings of instability and chaos. Releasing old fears calms the nervous system, preparing it for new impressions, while helping one to feel more centered and grounded in ones' new body.

Being weight bearing is especially important as we age. Allowing the weight to pass through the bones is essential to skeletal health. 65% of adults loss the movement in their hip sockets, resulting in hip injuries and other hip related problems. Releasing your psoas so the bones can bear weight, along with mobilizing the hip sockets, helps increase skeletal, visceral and hormonal health.

Women's Cycles

Releasing The Psoas Muscle Throughout Pregnancy

It is important to adapt the constructive rest position as the belly grows. An alternative position and one that is good for those who feel it is a strain to hold the legs up, is to find a chair or couch and lie down so that your legs can rest upon it (see Chapter Releasing the Psoas). As the belly grows and you can no longer rest on your back comfortably, it will be necessary to work in a partially reclined position, supported by pillows. It is important for the **spine to be supported evenly** from head to coccyx. As your pregnancy progresses, reduce the time you rest in the position to only five or ten minutes maximum. Trust your *inner voice*. Pregnancy is a time of growth - of many changes. It is a time of letting go and starting new. Because the weight of pregnancy helps free the psoas muscle, it can also be a time to stimulate a deeper inner sense of oneself; offering yourself and your baby a rich, nourished place to grow.

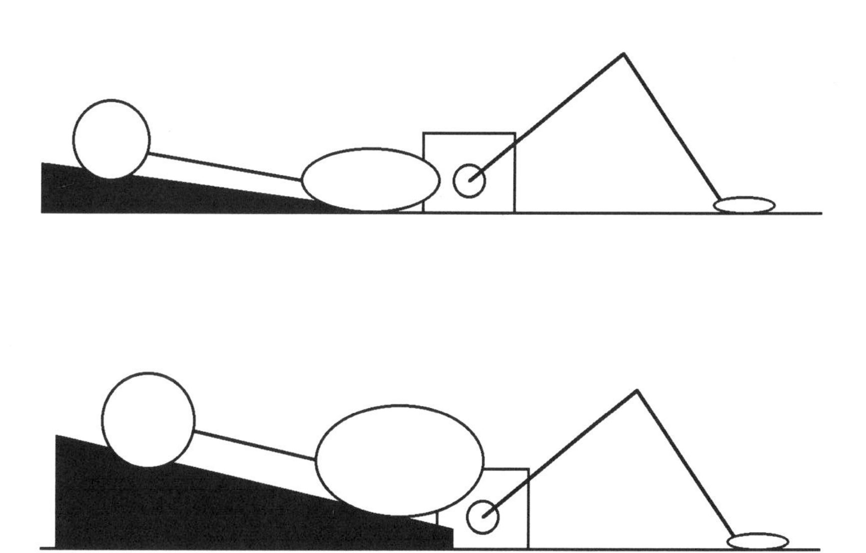

Application

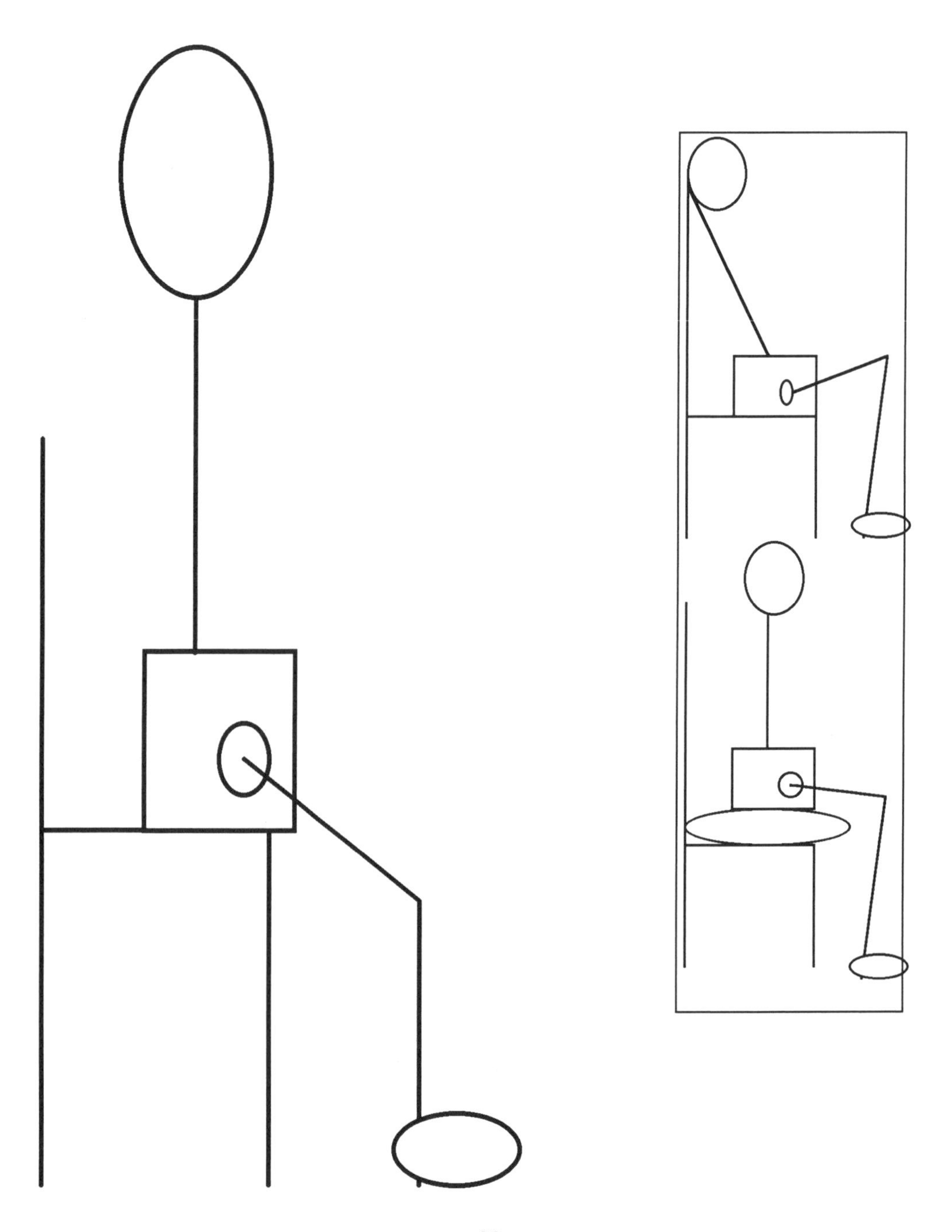

Chapter Eight
Applications

Shoes: Mobility in the foot is necessary for the sensitive anti-gravity nerves located in the heel and ball to fire properly. Shoes therefore play an important role in walking. Stiff soled shoes restrain the foot from rolling, pointed toed shoes restrict the toes from spreading, high heels forward trust the pelvis by tipping the center of gravity forward and stiff high tops stop the ankle from bending, limiting movement not only through the foot but all the way to the hip socket and effecting the quality of movement throughout the entire leg. The result is a lack of rhythm in walking, and the use of the psoas muscles to steady awkward movement.

A toddler's growing foot needs freedom to roll and move. Many children's shoes do not bend or flex. A shoe for a child needs to be comfortably wide, light and able to bend in half. It basically protects the foot from heat or cold, and rough surfaces. The older child or adult shoe also needs to bend in half, giving the toes room to move and not conform the foot (and leg) to a particular shape. Try bending your shoe in half, where is the flexibility that will allow your foot to roll? Take a look at the bottom of your shoe, you will see the sole of the shoe is also designed to shape your foot (and thus your leg) to a particular form. Select a shoe that is as **neutral** as possible and has only a small heel.

Chairs: Most chairs are not designed for sitting. Chairs are designed for stacking. When selecting a chair, notice the seat. Avoid bucket seats that form a hollow for your bottom. A chair needs to offer a solid base of support for the tuberosities to rest on. The next best choice is to modify the chair using a hard pillow or folded blanket. **KEY: your hip sockets must be higher than your knees for your psoas muscle to stay released.** Seats in a car create a more complex problem, because filling the bucket seat will lift your trunk up so high that your head may hit the ceiling; making visibility difficult. Obviously the best solution is to select a car that offers more control for shaping the drivers seat. The next best option is to modify the seat as best you can. Tilt the seat slightly forward and lift the back of your pelvis with a hard foam or rolled towel. Once again seek to open the hip sockets, freeing the psoas for responsive driving.

Desks: Sitting at a desk is also a complex situation. The height of the stool or chair must allow your pelvis to be stable and supported, freeing the psoas muscle and keeping the hip sockets released. The head and trunk need to be in relationship to each other for the upper psoas to release. The eyes look ahead rather than down. Modify the height of the table so you do not lean over it. Lift the monitor of a computer, for example, to a height that supports the head/neck righting reflexes, so important for stability. The height encourages you to look straight ahead when you type rather than down. Bending from the hip sockets (rather than at the waist or collapsing the head and shoulders) when you look down, reduces shoulder, neck and back fatigue. A constricted upper psoas instead of freeing the rib cage and elongating the spine, collapses the chest and hyper-extends the lumbar spine. Using a "back pillow" for the small of the back to relieve back pain (a result of a hyper-extended spine), is a bandaid approach to a much deeper problem, a constricted psoas.

People often ask about kneeling stools for sitting at desks. Kneeling stools do not allow the feet to be grounded (as the feet are not touching the floor). It is through the nerves in the bottom of the foot that we receive anti-gravity support. The body also grounds itself electrically (energetically) through the feet. Kneeling chairs increase stress in the knees, where tension already may be manifesting inappropriate torquing patterns.

Bikes: Children love to ride almost as soon as they can walk. Choose a little tike car that is pushed with the feet and allow the child to sit firmly on their pelvis. As the child grows the trike or bike seat needs to be adjusted to a height that keeps the pelvis stable and the hip sockets higher than the knee. The leg needs to be able to fully extend. Adjust the handle bars so the arms come straight from the shoulder to support the upper body.

Chapter Nine
Approaches

Sports, exercises, dance, yoga, movement arts and methods of body work may be seen in terms of how they approach and work with the psoas muscle. Literally each form creates a different body. Why we choose one form rather than another involves our conditioning, interests, understanding and goals.

Through activity our sensory perception and nervous system develop and mature. However, mechanically practicing movements neither develops our awareness nor matures the nervous system; instead we continue to deepen an already faulty way of moving. Repetition without increased awareness brings nothing new. To discover oneself from the inside out, we need to begin by learning to sense ourselves from the inside out.

Body Building and Weight Lifting: Body building and weight lifting are fashionable activities today. However releasing the psoas muscle is rarely a focus for learning to bear weight correctly. The emphasis instead, is on over developing external muscles which creates an armoring effect that can eventually distort the bony structure . It is the over developed musculature that actually torque's the bones and discourages them from bearing additional weight. To "work out" one needs to have a skeleton that is aligned and able to transfer weight efficiently and effectively. Otherwise the body attempts to compensate and problems arise (torn and bruised muscles, ligament strains, joint dysfunction).

Exercises and Sports: Exercising and various sport activities emphasize strength, endurance and speed. Development of muscle control rather than skeletal balance takes precedence. Gaining speed at the expense of mounting tension, is too often the goal. The psoas is rarely considered or engaged directly. However, it is possible to work from a released psoas muscle. First one must slow the body down until one can follow the *quality* of movement through ones sensory awareness. For example, if you wish to use your hip socket to rotate the leg, you will need to slow the movement down until you can sense yourself initiating the movement. You will need to mature your kinesthetic awareness. As you are able to follow your sensation you will notice for example that you initiate movement at the knee or the lumbar joints instead of the hip socket. One increases awareness by differentiating movement patterns in one's body. Sensing the old patterns comes first, then learning a new pattern. With increased awareness comes efficiency and speed.

Bicycling: Long distance bicycling will often over-develop leg muscles unless the body is in proper alignment and the bike is sized and adjusted to fit the person. Eventually the overuse of flexor muscles pulls the knee joint into inward rotation torquing the leg bones and putting pressure on the knee. Overdeveloped muscles stop a person from experiencing their legs as weight bearing. One question that must be asked is whether or not it is possible for a person to perform an activity repeatedly and not distort a particular muscle group? The human structure needs variety - engaging different muscles groups in different activities. However more importantly is the need to engage the right muscle group for any particular activity. Whenever one cannot perform an activity with the proper muscle group the body compensates by using alternative muscles. For example the overdeveloped use of the adductors (the muscles of the inner thigh) to counter the psoas muscle instead of the outward rotators (the opposing muscles in the hip), results in distortions and restrictions throughout the pelvis, leg, hip socket and pubis. A clue to the inappropriate use of the adductors appears when a person cannot experience the legs being stable and weight bearing while in the constructive rest position.

Running: Running like other activities does not necessarily in and of itself improve posture that is already poor and constricted. It often exaggerates problems due to the substitution of inappropriate muscles. Repetition in and of itself does not refine movement . When the psoas is already contracted it will be so during and after running. The diaphragm cannot function properly and do the work it is called upon to do. The repetitive inappropriate development of the musculature (as in body building or weight lifting) often leads to diminished sensitivity. Stress occurs in the knees and lower back, encouraging injury. Of course it is possible to engage the psoas properly while walking and running; but it involves learning to stand and walk without using the psoas as a structural support. Walking Instructor Suki Munsell, Ph.D. (see Resources) has developed a re-education program called *Dynamic Walking* whereby one learns to engage the psoas in active walking. The movement sequence of the leg is like a pendulum. Initiated at T12, it is the center of the trunk where walking begins - the leg simply follows. Proper effortless walking starts at the core - the psoas muscle. With each step the psoas moves through a full range of motion; releasing, stretching, neutral i.e. releasing and contracting. The legs move asymmetrically, each going through a alternating cross lateral movement pattern. This can only happen when the psoas is no longer used as a structural support system.

Swimming: Swimming is an activity that can either create structural problems or release them depending upon the way it is taught and practiced. Professional swimmers are known to develop shoulder tendonitis and kyphosis. Overriding head/neck righting reflexes (as occurs when the head is repeatedly turned but the body does not follow) eventually result in overdeveloping shoulder muscles, pinching nerves and distorting the rib cage. These are well-known problems in the field. Robert Cooley, Director of *The Moving Center Inc.*(see Resources) developed *The Swim Stroke* and presented it to the US. Olympics Advanced Coaches Seminar in Colorado in 1978. The swim stroke is based upon natural movement patterns of the arm and shoulder and not overriding the reflexes that coordinate these movements. The stroke engages the psoas muscle by allowing the whole trunk to arch freely from the hip socket, the leg follows. By freeing the psoas it releases and responds to each breath - the whole body swims. The natural movement includes leaving the head up and out of the water, the spine gently arches as the psoas releases and stretches. The arms can move in several natural patterns. Although the head responds to the movement of the arms, it does not rotate in a pattern of sideways rotation (i.e. with head locked and unresponsive to the arm movements) unless the whole body rolls to the side. Instead, the body is placed in the water, given the right ideas, and allowed to swim. The stroke is not imposed on the body but develops out of an awareness that keeps increasing the development of the motor skill.

Dance: We have seen that muscle control as a means of arranging posture results in muscle fatigue, loss of flexibility and inefficiency and yet most people approach dance by training their bodies to have strong muscular control. When dancers speak of the psoas muscle at all, the psoas is usually referred to as a muscle that needs strengthening. This is particularly true in ballet and related forms of dance in which the psoas is used to support weight rather than the bones being weight bearing and utilizing the psoas muscle to gain range of movement. *"In ballet, muscular contraction of the abdominal and gluteus maximus is intentional while work on the thigh is attempted. This is done to counteract the short resting length of the flexors and especially the iliopsoas muscles "(3)*

The trunk is held stable by contracting the upper psoas muscle while the legs are moved. To really strengthen the psoas in an appropriate way would mean releasing, toning and lengthening it , thus experiencing the bones bearing the weight and the psoas as a guide wire assisting skeletal balance. The over development of the large gluts (butt muscles) results in holding the pelvis hostage to the leg movements. It does not assist freeing the hip joint for full range (up to 160 degrees) range of motion.

A unique approach to dance is *The Hawkins Technique,* developed by the late Eric Hawkins (see Resources). This technique focuses on, and works directly with the psoas muscle. People who have worked with the Technique say they spend a lot of time learning to walk so that the leg "tracks" or moves properly at the hip socket.

Contact Improvisation, works with weight transference and bearing not only one's own weight but the weight of other people. To bear weight properly and fluidity, it is absolutely necessary to learn to release the psoas muscle. Weight then moves through the arms and spine, transferring at the hip sockets into the legs and feet without obstruction. The psoas is not spoken of directly, and yet there is an appreciation of the necessity of having a balanced weight-bearing skeleton and in turn releasing from the core any unnecessary holding in the musculature.

Movement Arts: Jujitsu, Aikido and especially Tai Chi, work in a way that is completely opposite from many forms of dance, specifically ballet. It seems that more and more people interested in dance and movement are turning to Eastern forms of movement as they search for richer and more supple expression.

In Tai Chi the body is placed in a position where the six outward rotators are eccentricity contracting with the abdominals and gluteals relaxed. This eccentric contraction of the six outward rotators counteracts the short resting length of the iliopsoas as well as gravity. Being in the tai chi posture utilizes gravity to one's advantage. The main difference then is in the use of the abdominals and the gluteals, and that in tai chi the force of gravity is utilized to stretch the iliopsoas and flexors, while in ballet gravity is not used. It is possible to use gravity to stretch the flexors and iliopsoas in ballet but this is not understood in the teaching of this art. (3)

Martial Arts students need a weight bearing skeleton, a free psoas muscle and open hip sockets to perform well. There is a tendency to lock the psoas muscle in a defensive posture and fatigue the muscle by keeping it in a cronic contracted state. This limits movement of the leg, encouraging the use of the lumbar spine for kicking and stops a person from performing effectively.

Yoga: A yoga practice encourages a limber, flexible body and mind. Yet once again how yoga postures are understood and approached, who is guiding the correction of postures, and what the ultimate intention is, will determine what is reaped. The yoga poses are exacting forms that if practiced precisely can stretch specific muscles groups, stimulate organ functioning, breathing and the meridian flows (energy pathways through the body). However what tends to happen, whether or not one is a student or a teacher, is an over extension of joints (that is a pulling apart of bones at the junctions) causing ligament stretching and potentially injuring the delicate joints.

What is not recognized is the need to keep the pelvis stable in each pose and to advance only as the pelvis and other bones stay in alignment. When the pelvis is the foundation of an asana, joints relate like pearls on a string. For one to gain not only flexibility but strength i.e. proper stretching an understanding of resistance in the position needs to be applied. Robert Cooley, Founder and Director of *The Moving Center* (see Resources) teaches proper stretching. He utilizes yoga postures for stretching in a form called - *Resistant Stretching* ™ and correlates each asana with its specific organ meridian - *The Meridian Stretching System*™ Based on his strong kinesthetic awareness and knowledge he teaches stretching always with the psoas muscle in mind.

Other teachers have incorporated their understanding of the psoas muscle by starting each class with releasing the psoas muscle. Resting in *the constructive rest position* before beginning even a "simple" pose, helps to release the body from many of the restrictions that limit ones' range of motion. To move into a pose, demands a stable pelvis and open released hip sockets.

Body Work

There are many varied and popular ways to align the body and sensitize one's awareness to movement. Each technique creates a different image as to what the properly aligned body should look like. Although there are some common grounds, each method will evolve a different body. For example, each approaches the positioning of the pelvis in a different way, therefore affecting how the psoas will be engaged. Judith Aston has written that " *when the angle of the pelvic crest is parallel to the ground (horizontal) the weight is placed on the heel - this is the model for Rolfing as well as for Sweigard and Alexander. What needs to happen is the resilient positioning of the pelvis onto the top of the legs; this would allow the chest to occupy more of its' depth , and the elbows, arms and head to realize a foot in a light manner. In this way, during motion, the shock absorbers of the body would accept the impact of weight allowing the body to actually massage itself while moving. (13)*

Rather than a Sport or Movement Art defining ones' body shape, it is possible to go from a neutral position into a specific activity and return once again to the neutral positioning. A neutral body that was not defined for example, as a "dancers body" or a "disk thrower's body". Instead the body would spiral into an activity and spiral back to a neutral aligned body.

It is difficult to know which body work method is best for any given person. However, looking at different methods in terms of the psoas sheds a certain light by which one may begin to see which methods are most suitable. The methods included are the ones which laid the foundation for many of the diverse combinations that have sprung forth in recent years, The ones chosen represent just a small handful of the varied forms of bodywork. The list does not include the varied forms of working with the *energy* fields. *Energy* work should not be underestimated in its direct ability to release and lengthen the psoas muscle. It is possible to release the psoas without ever holding points relating to the muscle or palpating the abdomen.

A form of work is only as good as the practitioner who is practicing it. The following descriptions are meant to be a guide and possibly a tool for selecting a form based on whether or not it works directly with the psoas muscle. Please remember that feeling safe is essential for allowing oneself to receive new impressions and release the psoas muscle. Ultimately the work must be done by you - it is your process. Another person can only assist and provide guidance.

The Alexander Technique ™ works with the idea that the "primary Control" located in the head/neck relationship is what controls posture

Working to release the head and neck, therefore, allows the spine torealign. It is a good idea and an accurate one and yet because people live too much in our heads, this is perhaps not the best place to begin to create change. the psoas is not a major focus of this techniques. But there is a great deal to be learned from understanding the righting reflexes in the head and neck.

Lulu Sweigard One of the leading people in the 1930s, Lulu Sweigard was concerned with what she termed the **"Human Movement Potential"** (14). Her and her coworker Mabel Todd, author of **The Thinking Body** (15), worked with images as a way of aligning posture. Sweigard and Todd emphasize the importance of the psoas muscle and their contribution to the bodywork field is very important. The approach,however (that of using imagery) and the many offspring's of their work, do not take into consideration that we are already holding images or ideas as to how to move. Adding new images or ideas does not necessarily correct or remove old ones. More importantly it does not bring one directly in contact with pure sensation. Exploring ones' body and listening to it speak, is a process of uncovering an ancient language. Too often we impose ideas, thoughts and images on our body; blocking the wisdom that will come forth if not hindered.

Structural Integration (Rolfing)™ Developed by Dr .Ida Rolf, the form acknowledges the psoas muscle as an important foundation for structural support. Dr. Rolf's ideas of posture are built around engaging the psoas properly. For many people the word psoas is synonymous with "rolfing". Rolfing has changed over the years. The technique is sometimes referred to as having softened. Whatever the style used, rolfing works with engaging the psoas by direct manipulation. Direct manipulation contradicts the function of the psoas muscle as a part of the fear reflex system. When the psoas experiences deep palpitation there is an instinctive response to contract. It is counterproductive to massage the psoas directly. It is an instinctive muscle and does not respond to intellectual messages. It is simpler and more empowering to voluntarily learn to release and sense the psoas oneself and/or with the guidance of a

---

trained therapist. Learning to voluntarily release your psoas muscle is an essential step to maturing the whole person and the conditioning that controls sensation and feelings. (see The Fear Reflex and Childhood Chapters) Photographs reveal the upper psoas has no facia surrounding it. It is a lean meat. Only the lower section around the hip joint has facia. Massage techniques that work with slowly releasing the hip socket can be advantageous. But most important is how quickly one opens to new impressions and how new impressions are integrated. Strong and powerful work can be overwhelming and will be immediately rejected by the nervous system. The location of the psoas is in our very gut where lives all our gut feelings. Subtle, non forceful and often less physical forms of work are just if not more digestible. There are non manipulative and non invasive techniques of releasing the psoas that are both effective and empowering for the person involved. Integration and changes happen at a pace and depth responsive to the person and in the context of their life. Ida Rolfs' contribution is respected and important and yet her model of a stacked body, although an informative image suggest a static three-dimensional form (like a house that needs a firm foundation), rather we need to see and experience our body as a fluid, moving changing organism.

Feldenkrais ™ Developed by Moshe Feldenkrais, Feldenkrais Work focuses on refining awareness as its' main tool for maturing the nervous system. The psoas is seen to affect the lumbar vertebrae, which are important in allowing the pelvis to be extended. when the pelvis is extended it creates a shelf from which the whole spine is supported (i.e. stable pelvis). Feldenkrais work involves the whole person in developing an awareness of him or herself. Although the right ideas are suggested, it does not impose an image. Emphasis is placed on the diagonal and lateral relationships; (that is on the relationships between the right and left side versus front to back). Volume so important to the 3-dimensional aspect of the body is not as emphasized as is cross diagonal patterns. Cross diagonal patterns are important in maturing the nervous system. We now know cross patterns help to "fire-up" the electo magnetic fields; energizing and balancing the whole person.

Approaches

Aston-Patterning ™ Developed by Judith Aston, Aston-Patterning emerged out of work Judith Aston started originally with Ida Rolf. Judith comes from a background that includes dance and movement. Her work evolved into focusing on the body in motion. It offers a three-dimensional quality to the body, which includes an awareness of volume and depth. This is the result of looking at the body as a form that continually moves in spirals. The body emerges not from an image of how it should look but from how it functions. Judith Aston speaks of the body unraveling as the persons' particular holding patterns release. She is looking not at linear relationships but full spiral/ diagonal relationships. The psoas muscle is an integral part of the unraveling. Although it is not verbally focused on, Aston-Patterning does release and engage the psoas; returning it to its' rightful place as a unifying factor in posture and movement. Aston-Patterning is the most sophisticated form of bodywork, integrating all aspects of the body/mind/emotions.

Chiropractic The term covers a wide range of techniques, philosophies and expertise. Craniopathy is a branch of healing that addresses the Cranial-Sacral Respiratory mechanism (skull and plevis) and its' importance in controlling the central nervous system. Craniopathy was first introduced in the late 1920s. The Vector Point System in particular has a profound means of changing the whole person, and re-establishing balance throughout all the systems of the body. There are Chiropractic avenues for correcting and healing unstable pelvis caused by ligament injuries. With the guidance of a trained therapist, A *trocanter belt* is worn to support the bones of the pelvis while the ligaments can heal. The trocanter belt maintains stability in the pelvis, freeing the psoas muscle fromholding the pelvis stable.

Be careful what therapist you choose. Chiropractic care that is based on continually adjusting joints results in increased irritation and a build up of synovial fluid. The problem grows into a cycle of need that does not address the issue of why the bones are out of alignment. When force is applied over time, ligaments stretch losing their tone and ability to hold the joints together properly. **Be an informed customer.**

Bibliography

1) Therese Bertherat and Carol Bernstein, *The Body has its' Reasons; Anti-Exercises and Self-Awareness,* Avon Books, N.Y. 1979 (first published in France 1976).

2) Susan K. Campbell, *Inherent Movement Patterns In Man, Kinesiology III,* 1973

3) Bob Cooley, *East-West Dance Forms,* The Moving Center, Inc. Boston, MA. Privately published paper, 1976.

4) Karlfried Graf Von Durkheim, *Hara, The Vital Centre of Man,* Mandala Book/Unwin Paperbacks, London, 197 (first published 1962).

5) Moshe Feldenkrais, *Awareness through Movement,* Harper & Row, N.Y. 1972.

6) Moshe Feldenkrais, *Body and Mature Behavior; a Study of Anxiety, Sex, Gravitation and Learning*, International Universities Press, N.Y. 1975 (first published 1949).

7) Ron Kurtz and Hector Prestera, MD. The Body Reveals; *An Illustrated Guide to the Psychology of the Body,* Harper & Row/Quicksilver Books, N.Y. 1976.

8) Michio Kushi, *The Book of Macrobiotics; the Universal Way of Health and Happiness,* Japan Publications, Inc. Tokyo, 1977.

9) Felix Mann M.B., *Acupuncture; The Ancient Chinese Art of Healing and How it works Scientifically,* Vintage books, N.Y. 1973.

10) Willhelm Reich, *The Function of the Orgasm*, Noonday Press, N.Y. 1970 (first published 1949)

11) Ida P. Rolf, Ph.D. *Rolfing - The Integration of the Human Structure*, Harper & Row, N.Y. 1977.

12) Rosemary Feitis,ed. *Ida Rolf Talks About Rolfing & Physical Reality,* Harper & Row, NY. 1978.

13) *"A Somatics Interview with Judith Aston"*, from Somatics Magazine, privately reproduced brochure.

14) Lulu Sweigard, *Human Movement Potential; Its Ideokinetic Facilitation,* Harper & Row N.Y. 1974.

15) Mabel Elswoth Todd, *The Thinking Body; A study of the Balancing forces of Dynamic Man,* Dance Horizons, Inc. N.Y. 1978 (first published 1937).

Resources

- **Aston-Patterning** offers private sessions & teacher training. Sessions include educational as well as hands on work Contact Judith Aston @ Aston-Patterning, P.O. Box 3568 Incline Village Nevada 89450 (702) 831-8228 for a Patterning Teacher in your area.• **Ronnie Oliver** , Aston-Patterning practitioner & teacher in San Francisco, CA., Certified by Judith Aston in '78 and on the Faculty of the Aston Training Center (U.S., Canada & New Zealand), may be reached at (415) 648-1718

- **Dynamic Fitness and Health Institute** offers postural workshops, body exercises and dynamic movement training for every level of participant. Instructor & professional Training's also available. *Dynamic Movement* was researched and developed 25 years ago by Suki Munsell Ph.D. It is taught nationwide by certified instructors in hospitals, health & fitness center, corporations retreat and universities. **Contact Dynamic Health & Fitness, P.O. Box 355 Corte Madera, CA 94976 (415) 924-4013 • Fax (415) 924-5342 • email: dynamic@well.com**

- **Erick Hawkins Dance Foundation Inc.** School and foundation located at 375 W. Broadway, New York, N.Y. (212) 501-2529

- **The Moving Center Inc.** offers classes, private sessions and instructor training in *The Meridian Stretching System™* and *Resistant Stretching™.* For more than 20 years Founder and Director Robert Cooley has been exploring kinesthetic awareness. The Moving Center is a learning center focusing on understanding the human body in motion. Always changing, dynamic, and profound, Bob Cooleys' work is the state of the art in new ways of thinking about the human body. Contact The Moving Center Inc. 29 Commonwealth Ave. Boston, MA. (617) 437-YOGA

- **Vector Point Cranial-Sacral Technique** Victor Collins Doctor of Chiropractic's works with several Sacro-Occipital Techniques, Dr, Denton's Vector Point System. Recommended for tocanter belt diagnosis: 526 Soquel Ave. Suite D Santa Cruz, CA (408) 458-1231

Resources

Liz Koch - Author of The Psoas Book, Liz Koch taught workshops and classes focusing on the psoas muscle and its profound affect on the body/mind/emotions, for over 20 years throughout the United States Her teaching experience includes: North American College of Natural Health Sciences, Holistic Life University, Mills College Graduate Dance Program, University of California at Santa Cruz, Tufts University, and The Boston Museum School of Fine Arts where as an instructor of the sculpture faculty she directed the *Flexibles* Program, integrating sculpture and movement. Liz is a certified Jin shin Do Acupressure previously on the staff of the Alternative Therapies Unit at San Francisco Hospital. Liz studied Jin Shin Do with Aminah Raheem and Iona Teeguarden and Kinesthetic Awareness with Robert Cooley Director of The Moving Center Inc. Boston Massachusetts.

Workshops: focus on learning to release, tone and lengthen the iliopsoas muscle. If you would like to attend, sponsor or have a class /workshop designed for your teaching program contact Liz Koch @ Guinea Pig Publications or visit her web site at www.coreawareness.com

Books: **Individual Orders** – Retail $22.95 + Shipping Priority $3.95 (1-4 books going to the same address. CA Residents add your local sales tax or 7.75% to total sale.

SIMPLIFY ORDER: $28.00 (CA Order) $26.90 (Out of State)

International Orders: Send bank money order drawn on US Funds Shipping Global Priority is $7.00 Canada (1-4 books) $9.00 Other

Bulk & Wholesale Orders: see order form included or write orders@guineapigpub.com for current prices.

Web Sites: www.guineapigpub.com www.coreawareness.com

Guinea Pig Publications P.O. Box 1226 Felton CA 95018

Order Form

Guinea Pig Publications P. O Box 1226 Felton CA 95018
(831) 335-1851   Orders@Guineapigpub.com

☐ Please send #_____copies of ***The Psoas Book*** $22.95 ea _____
☐ Add $3.95 Shipping in USA (for 1- 4 books delivered to the same address) **$3.95 ***
(CA Orders must add shipping fee before calculating tax) SUBTOTAL _____
☐ CA Residents must add your local California Sales Tax or 7.75% _____
☐ Please send #_____ copies of Articles **listed below @ $3.50 Each** _____
TOTAL ENCLOSED: __________

Copy Quality Articles by Liz Koch $3.50 EACH (USA) $5.00 (International)

☐ ***Pregnancy/Birth and The Psoas Muscle*** (printed in *The Doula Magazine)*
☐ ***Refuse to Fuse***: *Adolescent **Scoliosis*** (privately published)
☐ ***Your Back In Gardening*** (versions printed in *Vegetarian Times & Connections*)
☐ ***Core Awareness: 5 Most Important Things to Know about your Psoas Muscle***
(Printed in *Yoga & Health* (London England)

INTERNATIONAL SHIPPING ORDERS FOR: ***The Psoas Book***
Please send bank money order drawn on US funds.
Add Shipping fee of ***CANADA $7.00 *EUROPE /Other Countries $9.00**

BULK & WHOLESALE ORDERS; Contact orders@guineapigpub.com

Please Print Clearly

Name____
(As it appears on credit card)
Street____
(Please include the credit card's billing address if different than mailing address)

City____________________ State______ Zip_______ Tele:________________

Type of Credit Card______ #__________ __________ ________ ________

Expiration Date__________ Signature______________________________
Total Payment Enclosed $__________
Please include me on your mailing list for workshop dates & information

☐ YES ☐ NO

Make Check or Money Order Payable to Guinea Pig Publications

Order Form

Guinea Pig Publications P. O Box 1226 Felton CA 95018
(831) 335-1851 Orders@Guineapigpub.com

☐ Please send #_____copies of ***The Psoas Book*** $22.95 ea _____
☐ Add $3.95 Shipping in USA (for 1- 4 books delivered to the same address) **$3.95 ***
(CA Orders must add shipping fee before calculating tax) SUBTOTAL _____
☐ CA Residents must add your local California Sales Tax or 7.75% _____
☐ Please send #_____ copies of Articles **listed below @ $3.50 Each** _____
TOTAL ENCLOSED: __________

Copy Quality Articles by Liz Koch $3.50 EACH (USA) $5.00 (International)

☐ ***Pregnancy/Birth and The Psoas Muscle*** (printed in *The Doula Magazine)*
☐ ***Refuse to Fuse***: *Adolescent **Scoliosis*** (privately published)
☐ ***Your Back In Gardening*** (versions printed in *Vegetarian Times & Connections*)
☐ ***Core Awareness: 5 Most Important Things to Know about your Psoas Muscle***
(Printed in *Yoga & Health* (London England)

INTERNATIONAL SHIPPING ORDERS FOR: *The Psoas Book*
Please send bank money order drawn on US funds.
Add Shipping fee of ***CANADA $7.00 *EUROPE /Other Countries $9.00**

BULK & WHOLESALE ORDERS; Contact orders@guineapigpub.com

Please Print Clearly

Name____
(As it appears on credit card)
Street____
(Please include the credit card's billing address if different than mailing address)

City____________________ State______ Zip_______ Tele:________________

Type of Credit Card______ #__________ __________ ________ _______

Expiration Date_________ Signature_________________________________
Total Payment Enclosed $_________
Please include me on your mailing list for workshop dates & information

☐ YES ☐ NO

Make Check or Money Order Payable to Guinea Pig Publications

NOTES

NOTES

NOTES

NOTES

The Psoas Book

A Comprehensive Guide to The Iliopsoas Muscle and It's Profound Effect on the Body/Mind/Emotions

*"**The Psoas Book** expanded my vision of the functioning psoas"*
Janis G. Davis MD. (Rolfer for 25 years) St. Petersburg, Florida

"We love ***The Psoas Book***"
Jivamukti Yoga Center, New York, NY.

*"**The Psoas Book** will teach you what the psoas muscle does physiologically, why it gets stressed and how you can relax the muscle in order to restore optimal health"*
Dr. Victor Collins.D.C. Santa Cruz, CA

*"**The Psoas Book** is required reading for my Teaching, Terminology and Technique Classes"*
Joanna Mendl Shaw, Dance Instructor (Seattle, Washington)

*"**The Psoas Book** is a must read for everyone interested in Conscious Fitness"*
Suki Munsell, Ph.D., Dynamic Health and Fitness Institute, Corte Madera, California

Price $22.95 ISBN 0-9657944-0-7